Repair, Protection and Waterproofing of Concrete Structures

Repair, Protection and Waterproofing of Concrete Structures

Third edition

P.H. Perkins

CRC Press
Taylor & Francis Group
Boca Raton London New York

CRC Press is an imprint of the
Taylor & Francis Group, an **informa** business
A TAYLOR & FRANCIS BOOK

CRC Press
Taylor & Francis Group
6000 Broken Sound Parkway NW, Suite 300
Boca Raton, FL 33487-2742

First edition 1977
Second edition 1986
Third edition 1997

First issued in paperback 2019

CRC Press is an imprint of Taylor & Francis Group, an Informa business

Typeset in 10/12pt Palatino by Mews Photosetting, Beckenham, Kent

ISBN-13: 978-0-419-20280-6 (hbk)
ISBN-13: 978-0-367-86639-6 (pbk)

A catalogue record for this book is available from the British Library

Library of Congress Catalog Card Number: 97-066019

Contents

Preface to the First Edition

While there are examples of reinforced concrete structures built towards the end of the nineteenth century and the early part of the twentieth, concrete as a general constructional material only began to be used on a larger scale after the end of the First World War. The military requirements of the Second World War and the large development and rebuilding programme which followed it, established concrete as the major constructional material. Structures erected in the 1920s and 1930s are now between 40 and 50 years old, which will be within the memory of many readers of this book.

The use of a relatively new building material inevitably brings problems which were not anticipated initially and disappointments are common. Reinforced concrete is no exception and due to the increasing age of the early structures, the need for repair and renovation is increasing.

While the structure and architectural design of buildings varies widely from one country to another, the principles of repair are more universally applicable. Therefore the author hopes that the contents of the book will be useful to a very wide range of persons who are responsible for the maintenance of concrete structures of all types.

The opinions and recommendations in his book are those of the author, but he is indebted to his colleagues in the Cement and Concrete Association and to staff in the leading firms which specialise in the repair, protection and waterproofing of all types of concrete structures.To all these people the author expresses his sincere thanks. He also wishes to thank the staff of the Cement and Concrete Associations of Australia, New Zealand and South Africa for their help in compiling the relevant sections in Appendices 1 and 3.

Philip H. Perkins
1976

Preface to the Second Edition

Since the author's first book on the repair, waterproofing and protection of concrete structures was published in 1976, the need for repairs to this type of structure has increased dramatically.

It has been estimated that in 1982, the value of repairs and maintenance to buildings in the UK amounted to some £8 × 10^9 and if civil engineering structures were included this would rise to £1 × 10^{10} (£10 000 million). These figures relate to all types of buildings and structures of which concrete forms only a part. The disturbing feature is that in most cases the structures are not more than about 25 years old. The problem is not confined to the United Kingdom.

While the structural design of reinforced concrete varies to some extent from one country to another, the materials used, Portland cement, aggregates and steel reinforcement, are essentially similar. The causes underlying the deterioration are basically the same in all countries and the principles involved in dealing with the deterioration are also similar.

This book is intended to deal mainly with 'non-structural' repairs, that is repairs which are intended to restore long-term durability, but which will not increase to any significant degree the load bearing capacity of the structure. Mention is made of certain aspects of the execution of structural repairs but the calculations necessary for the design are not included. However, one of the first and, most important steps in the investigation of a deteriorated structure is to decide whether structural strengthening of the structure is needed. For this reason the author recommends that all such investigations should be carried out by a Chartered Civil or Structural Engineer with considerable experience in this type of work. It should be remembered that the investigation and diagnosis and subsequent preparation of specification for remedial work to a deteriorated structure is quite different to the design of a new structure.

Reinforced concrete structures which have been properly designed and constructed and which operate under normal conditions of exposure and use, require only a minimum of maintenance. However, it is a fundamental error to assume that they are maintenance free. Regular and careful inspection and maintenance are essential for all structures.

It is important that the lessons which can be learnt from the correct diagnosis of the present problems should be put to practical use in the design and construction of new structures.

The author wishes to express his gratitude to the many professional men from whom he received help and information. Special thanks are due to Keith Green of Burks, Green & Partners, Consulting Engineers, and George Korab of the Cement Gun Group of Companies for many informative discussions. Mention must also be made of the work of Lucy Perkins for reading and checking the manuscript.

Philip H. Perkins
1986

Preface to the Third Edition

Since the publication of the Second Edition of this book in 1986, the need to carry out extensive repairs to reinforced concrete structures has continued. Most of these structures were originally expected to have a life span of 80–100 years and this premature deterioration has caused serious concern.

In this country considerable publicity has been given in the technical press to major repair work on road bridge structures, particularly on major trunk roads and motorways.

While structural design, including specification, varies to some extent from one country to another, the materials used – namely Portland cement, aggregates, steel reinforcement and mixing water – are essentially similar. The main cause underlying this deterioration is also basically similar, namely the corrosion of the steel reinforcement. The journal, *Building and Civil Engineering-Research Focus*, April 1995, p. 5, says:

> The corrosion of steel reinforcement is the most serious durability problem affecting concrete structures throughout the world ... A possible alternative solution is the use of non-ferrous fibre reinforced plastics (FRP). EUROCRETE is a three-year £4 million EUREKA project with partners in the UK, Netherlands, France, Switzerland and Norway ...

Repair methods show somewhat wider differences. For example, in the US and Canada, cathodic protection has been used for many years before it was given serious consideration in the UK. The removal of chlorides by electrochemical means has also been tried in the US, but with mixed success. In France, considerable use (appreciably more so than in the UK) is made of special elastomeric coatings to increase the durability of the repaired areas of concrete, and to reduce greatly the risk of corrosion of reinforcement in new construction. These coatings are formulated to reduce water penetration into the concrete and comparatively little attention is given to the ability of the coating to prevent/reduce diffusion of carbon dioxide into the concrete.

This book is intended to deal mainly with 'non-structural' repairs, that is, repairs which are intended to restore as far as practicable, long-term durability and 'useful life' of the structure or part of the structure.

The expressions 'long-term durability' and 'useful life' are almost impossible to define in a clear-cut way, and therefore in Chapter 3 I have included some comments on this subject.

Non-structural repairs will not increase to any significant degree the load-bearing capacity of the structure.

One of the first and most important steps in the investigation of a deteriorated structure is to decide whether structural strengthening is required and, if so, whether the result is likely to be cost-effective.

I recommend that investigations of deteriorated reinforced concrete structures should be carried out by a chartered civil or structural Engineer, or other professional with considerable experience in this type of work. It should be remembered that the investigation, diagnosis and subsequent specification for remedial work are quite different to the design of a new structure.

I have included in Chapter 1 certain aspects of the investigation and repair of structures which my experience suggests are relevant and important to those associated with this type of work. While mention is frequently made to the engineer, especially his or her duties and responsibilities, the term is intended to apply to architects, building engineers, contractors and others who have responsibilities for the repair and renovation of concrete structures.

I wish to express my thanks to my professional colleagues, architects, engineers and specialist contractors, for help, information and advice in the revision of this book. Mention must also be made of the help and encouragement given by my wife, Lucy Perkins.

Philip H. Perkins
1997

Note: In the text of this book, where references to engineers, architects, contractors and other building personnel are given as 'he', this has been done to avoid the rather cumbersome repetition of 'he and she', and the author and publisher do not wish to imply that engineers and other building personnel are only male.

1

General observations

1.1 INTRODUCTION

While reinforced concrete structures which have been properly designed and constructed are resistant to deterioration, they should not be considered as maintenance free. Regular and careful inspection and the implementation of a sensible maintenance programme are essential. Reference can usefully be made to BS 8210: Guide to Building Maintenance and Management. 'Maintenance', 'useful life' and 'design life' are commented on in the Introduction to Chapter 3.

It is important that the lessons learned by the investigation of deteriorated structures should be taken into account when designing and constructing new structures. These 'errors' generally arise from inadequate and poorly drafted specifications and poor workmanship on site, but very seldom consist of structural inadequacy arising from errors in original design.

1.2 THE RESPONSIBILITIES OF THE ENGINEER OR OTHER PROFESSIONALS

An engineer who is instructed to investigate and report on a deteriorated concrete structure, and to prepare recommendations for necessary remedial work, should be clear in his own mind on the extent of his responsibilities to his client.

If he is responsible for the inspection of the remedial work to ensure that the requirements of the contract are complied with and the certification of payments-on-account to the contractor, then it is in the interests of his client that good relations with the contractor are maintained. These 'good relations' will be reflected in the standard of work, and its completion within the contract period.

The use of proprietary methods of repair can introduce problems in clearly defining the responsibilities of the engineer. For example, cathodic protection and, to a lesser extent, realkalization of concrete have come into use in the UK in the last few years. Both repair systems are highly specialized and unless the engineer happens to be well experienced in the use of such system(s), he would be well advised to make it clear in writing to his client that while he will accept responsibility for the integration of the specialist system(s) with other work in the contract, he will not be responsible for the efficacy of such systems, which must then be the contractural responsibility of the specialist firms concerned. The engineer's fees should clearly reflect this 'opt-out'.

The contract would then have to be drafted in such a way that the specialist contractor would be responsible for both the design and execution of the sections of the specialist work covered by the contract. This may sound rather academic to those who have not had experience with the law, but I refer to the High Court Case No. 1980-P-1364, known as *Pirelli v. Oscar Faber*. The Judgment by His Honour Judge Stabb, QC is dated 1 August 1980, and consists of some 32 pages, including the following statement which is relevant to the point being emphasized here:

> consulting engineers ... were not entitled to divest themselves of the duty of design entrusted to them unless expressly so agreed by their client.

1.3 BASIC PROCEDURE FOR INVESTIGATIONS – LITIGATION NOT INVOLVED

In cases where litigation is not contemplated, the basic procedure recommended to be adopted by the engineer following his appointment by the building owner is set out in Chapter 4 which deals with investigations and diagnosis of defects in reinforced concrete structures.

1.4 PROCEDURE WHEN LITIGATION IS CONTEMPLATED

The amount of litigation arising from the need to repair defects in reinforced concrete buildings has increased considerably in the last 20 years. This is reflected in the very high premiums charged by insurance companies for professional indemnity policies. The very real possibility of litigation arising from the need to carry out remedial work to existing structures must be taken into account by an engineer instructed to investigate a deteriorated structure. The engineer should realize that his report may well be used as the basis for legal action as the wording of the report will contain an opinion on the cause of the defects, which would indicate where responsibility probably lay.

Should the engineer's client initiate legal action to recover the cost of the remedial work, the engineer would normally be required to give evidence in court or before an arbitrator. Some comments on the engineer as an expert witness are given in section 1.5.

My experience suggests that it is always prudent to enquire at the time of appointment whether litigation is likely to be contemplated. The answer will influence the conduct of the investigation and the wording of the report. When litigation is contemplated or is already under way, the suggested approach is outlined below.

The engineer should, with the agreement of his client, ensure the following.

1. That all interested parties have been informed of his appointment and the reasons for it;
2. He should seek to reach agreement with the parties concerned on the details of the sampling and testing he proposes to carry out, and the testing laboratory he proposes to use. This is common sense, but is frequently neglected. Clause 4.2 in BS 60890: 1981: *Assessment of Concrete Strength in Existing Structures,* states:

   > Before any programme is commenced, it is desirable that there is complete agreement between the interested Parties on the validity of the proposed testing procedure, criteria for acceptance, and the appointment of a person or laboratory to take responsibility for the testing.

   The idea that there may be 'complete agreement' is in my experience, over-optimistic and is seldom achieved.

   While the Standard quoted only refers to the investigation of concrete strength in existing buildings, the principles are valid for all investigations involving the likelihood of a serious dispute.
3. There should be complete openness on the reasons for the investigation; all reasonable steps should be taken to avoid confrontation as this invariably results in a hardening of attitudes and resistance to objective discussion.

   Should it become clear that the only way to resolve the dispute in such a way that the building owner obtains reasonable compensation for the rectification of the defects in the building, then 'some form of legal action' is likely. 'Some form of legal action' means arbitration under the Arbitration Acts or the issue of a Writ in the High Court.

It is wrong to believe that arbitration will cost any less than an action in the High Court. In addition to the normal legal costs of solicitors' and counsel's fees, the arbitrator charges for his services and there is also the cost of hiring suitable accommodation for the hearings.

Most construction contracts include a compulsory arbitration clause, but the court can overrule this requirement. Also, no third party proceedings are permitted in arbitration.

In court proceedings, difficult technical considerations can arise if a defence of 'Limitation' is put forward. This would involve the Limitation Act 1980 and the Latent Damage Act 1984, and the engineer may be asked for his opinion on the following two issues:

1. When did significant damage first occur?
2. What was the earliest date on which the Plaintiff had both the knowledge required to bring an action for damages in respect of the relevant damage, and a right to bring such an action?

My experience is that such questions give rise to very complex technical considerations to which there is seldom a clear-cut answer.

Due to the enormous cost of High Court actions, proposals have been made in recent years to find alternative methods of settling disputes. This is generally known as 'Alternative Dispute Resolution'. Even when the claim is for hundreds of thousands of pounds, the cost of taking the dispute to court can be disproportionate to the possible financial benefit. While the majority of cases in the Official Referee's courts and in arbitration settle before trial, few do so early enough to avoid substantial costs incurred in the preparations leading up to trial.

The essence of ADR is to create a framework in which the parties involved in a dispute can reach a solution for themselves. This usually requires the assistance of a neutral third party.

There are a number of ADR techniques, such as:

- conciliation;
- mediation;
- mini trial;
- expert fact finding and adjudication.

The success of ADR depends entirely on the willingness of all the parties to resolve their dispute in a mutually satisfactory way, and this requires considerable give and take. Some references on ADR are included in the 'Further reading' section at the end of this chapter.

1.5 THE ENGINEER AS AN EXPERT WITNESS

It is possible that an engineer engaged by a building owner to investigate defects in a concrete structure will, as a consequence of such an investigation, be requested to act as an expert witness in any subsequent legal proceedings arising out of the investigation. This is different to acting as a witness of fact and requires a different approach, and is a complex subject on which quite a lot has been written. The following are extracts from four High Court judgments which highlight important aspects of expert evidence.

The first deals with the admissibility of expert evidence; the second with the necessity for the evidence to be objective and unbiased; the third with the liability of the expert to his client; and the fourth with the rejection of the expert's evidence on the grounds of partiality and bias.

1. *The Times*, Law Report, 17 July 1995. Lord Justice Stuart-Smith said that the admissibility of expert evidence was governed by Section 3 of the Civil Evidence Act 1972 and rules of court. An expert was only qualified to give evidence that was relevant if his knowledge and expertise was beyond that of a layman and such evidence had to relate to a factual issue in the case. The opinion of such witnesses could not consist of conclusions in respect of findings of fact which were strictly matters for the trial Judge to determine
(House of Lords, *Liddell v. Middleton*).

2. *The Times*, Law Report, 5 March 1993; 'The Ikarian Reefer' (1993) FSR 563. Mr Justice Cresswell said:

   Expert evidence presented to the Court should be, and should be seen to be, the independent product of the expert, uninfluenced as to form or content by the exigencies of litigation. An expert witness should provide independent assistance to the court by way of objective unbiassed opinion, in relation to matters within his expertise. He should not omit to consider material facts which could detract from his concluded opinion.

3. *The Times*, Law Report, 11 November 1991: *Palmer and Another v. Durnford Ford (a firm) and Another*. Mr. Simon Tuckey QC (Judgment, 31 October).

   An expert witness could not claim immunity from suit by his clients for his actions in the course of preparing evidence for a claim or a possible claim.

   Mr. Simon Tuckey QC so held in a judgment delivered in open court.

4. The following judgement illustrates very clearly what an expert witness should **not** do:

   Construction Industry Law Letter, September 1995, judgment reported: *Cala Homes v. Alfred McAlpine*. Chancery Division, Laddie J. Judgment delivered 6 July 1995.
   In August 1990, the expert witness had written an article in the journal of a professional institution, entitled, 'The Expert Witness; Partisan with a Conscience'. Issues and Findings in the case included:

   What was the relevance of the article by the expert five years ago?

The article, which the expert stood by in evidence, showed that he allowed himself to be unduly partisan. His evidence should be disregarded.

It is felt in some quarters that the above comment was rather too harsh.

The above extracts show that the giving of expert evidence is not a simple matter and there is suspicion that the experts for the various parties are not always impartial. Relevant to this is the inquiry being conducted by Lord Woolf to review the rules and procedures of the civil courts of England and Wales. The 27 chapters of his interim report deals with various matters and one of these chapters is devoted to experts. The comments provoked considerable controversy; the main points made were:

- The judge should identify the key issues in the dispute and he should decide what expert evidence will be needed to help decide them.
- The judge should have the power, with or without the agreement of the parties, to appoint a court expert to report and give evidence on these issues.
- The expert's report should be addressed to the court. Once an expert has been instructed to prepare a report for use of the court, any communication between him and the client or the client's advisers should no longer be subject of legal privilege.

The above is a very brief outline of Lord Woolf's interim proposals.[1]

At the present time the engineer may be appointed direct by his client to give a technical opinion on a dispute which had arisen. Should the dispute eventually end up in court, then all correspondence prior to the client's seeking legal advice is generally required to be disclosed to the other parties. It is therefore prudent for the engineer to endorse reports and comments issued in this period with the words 'Prepared with a View to Litigation', as this may then be claimed by his client's solicitors as 'privileged' and if this is upheld by the court these documents would not be available to the other parties.

However, the appointment (instruction) of expert witnesses is usually made after litigation has started and often well advanced and generally by the solicitors to the parties involved, namely, the plaintiff, defendant, third party etc.

It is usual practice in England, but not in Scotland and N. Ireland for the judge to include in his directions, an order that the experts' reports should be exchanged by a fixed date prior to the hearing. Before the reports are exchanged the experts are also required to meet for a without-prejudice discussion to try to find specific points on which they can agree. Specific points on which agreement is not reached are also sometimes recorded. I have found such meetings both interesting and useful, even though little or no agreement is reached.

1.6 THE PREPARATION OF SPECIFICATIONS

There are two basic types of specification. The 'traditional' specification which details materials and workmanship, and the 'performance' specification which details how the finished product must behave under clearly defined operating conditions and leaves how this is to be achieved to the contractor. The appearance of the product is often referred to by comparison with agreed samples.

An excellent example of a comprehensive performance specification is the one produced by the PSA for Platform Floors (Raised Access Floors) reference number MOB.PF2.PS a document of some 92 pages. A much more simple one is the Designed Mix, specified in Section 2 of BS 5328: Part 2: 1991 – *Concrete: Methods for Specifying Concrete Mixes*.

Both types of specification leave open the problem of long-term durability under the conditions of exposure and use which were originally contemplated at the design stage. Proper maintenance is essential for long-term satisfactory performance and this is discussed in Chapter 4.

1.7 THE CONTRACT DOCUMENTS

The type of documents which the engineer should prepare will depend on whether the client has a standard format, and the size and complexity of the repairs required. Subject to this, I favour basic simplicity and have found that the following is usually adequate for smaller jobs:

- Instructions to Tenderers
- Description of Works
- Specification
- Schedule of Work (each item of which has to be priced by the tenderers)
- Schedule of Rates (not always needed)
- General Conditions of Contract (simplified)

It is advisable for the Schedule of Work to include an item for the provision of access equipment (scaffolding, cradles etc.). Such equipment is usually hired by the contractor from specialist firms. The contract documents should make it clear that the contractor is responsible for ensuring that this equipment is fully utilized and claims for additional hire charges would only be considered if this resulted from an authorized extension of time. This is important because I have seen many jobs where scaffolding has been in place for weeks without any work being carried out from it and found out that in the majority of cases the client has had to pay.

1.8 INVITATIONS TO TENDER

There are many different ways of inviting tenders for a project. The one usually adopted for repair work is on a list of selected contractors. Depending on the type of work required to be done, the contractors invited to tender should be, and normally are, firms which specialize in the type of work described in the contract documents.

In some cases the specialist contractors put forward their own proposals which often include the use of proprietary repair materials such as prepacked mortars (ready to use apart from the addition of a gauging liquid). On the other hand the engineer may require the use of proprietary mortars which would then be referred to by name in the specification. These mortars are sometimes referred to as 'repair systems'.

The main problem in 'going out to tender' is the adjudication of the tenders in consultation with the client. The usual practice is for the client to expect the engineer to recommend the lowest tender, on the basis that the contractors tendering were selected by the engineer. The acceptance of the lowest tender can produce unfortunate results, especially when work is in short supply and contractors are naturally very anxious to be successful in tendering.

It is not always advisable to accept the lowest tender, as this may impose considerable financial strain on a contractor which has underpriced the job.

When there is a significant difference in the tender figures, the tenders should be carefully scrutinized with particular attention to the prices in the Schedule of Work (or Bills of Quantities), as well as any comments made by the tenderers. The prudent solution is usually to recommend the middle-priced tender. The experienced engineer and enlightened client will accept that the contractor is entitled to make a reasonable profit and if he can do so this will help ensure a satisfactory job which will be completed on time.

The following is an extract from an article on a large (£29 million) commercial development in the City of London which involved many technical problems of design and construction.[2]

> Both consultant and contractor found that an enlightened attitude by the client was a great help in overcoming these problems ... A good working atmosphere was achieved by two means; all companies involved had to sign an agreement known as The Accord, an undertaking to do everything possible to avoid conflict.

> The other measure, which probably was more effective, was the decision to provide extra money for subcontractor packages ... It enabled Laing to appoint specialist subcontractors by reputation rather than by just accepting the lowest bid. By working it this way, the client knew the final cost from the outset, rather than having a low bid followed by a lot of claims at the end of the work.

1.9 INSURANCE-BACKED GUARANTEES AND WARRANTEES

In recent years there have appeared on the market 'insurance backed guarantees'. These are offered by contractors and material suppliers claiming that should the work prove defective, then the insurance company will provide the funds to have the necessary remedial work put in hand in the event of the contractor/supplier failing to do so. This indicates that the client avoids resorting to legal action. These 'guarantee's', which are sometimes referred to as 'warrantees' are stated to be valid for periods of 10–20 years from completion of the work.

I have had occasion to see a number of these guarantees/warrantees, and found that they contained many anomalies and uncertainties. The conclusion is that such documents should be scrutinized by a solicitor experienced in that particular field. An engineer would be unwise to recommend reliance on such a guarantee without first taking competent legal advice.

1.10 NATIONAL AND EUROPEAN STANDARDS AND CODES OF PRACTICE

Frequent reference is made to British Standards and Codes of Practice. The principal British Codes and Standards relating to design and construction of reinforced concrete buildings and structures covered by this book are:

- BS 5400: Parts: 1, 2, 4, 7, 8, 9 and 10: Steel, Concrete and Composite Bridges
- BS 6349: Parts 1–7: Maritime Structures
- BS 8007: Code of Practice for Design of Concrete Structures for Retaining Aqueous Liquids
- BS 8110: The Structural Use of Concrete, Parts 1, 2 and 3
- BS 5328: Concrete, Parts 1–4

The corresponding Euro Codes for BS 8110 and BS 5328 are:

- Euro Code 2: Design of Concrete Structures: DD ENV2 1992: General Rules and Rules for Building

- DD ENV 206: 1992 Concrete: Performance, Production, Placing and Compliance Criteria (implemented in the UK). There is also a 'Guide' to the Code published by BSI as PD.6534: 1993.

It should be noted that European Standards are being drafted for products and systems for the protection and repair of concrete construction works.

The following terminology is in general use:

DD	Draft for Development
PD	Published Document
CEN	European Committee for Standardisation
CPD	Construction Products Directive
CE	Symbol for CE conformity Marking
EN	European Standard
pr EN	Draft European Standard
ENV	European Prestandard
prENV	Draft European Prestandard
ISO	International Organisation for Standardisation
CEN/BTS I	Particular Sector Technical Board which manages the Standards work programme
SCC	Standing Committee on Construction
TCs	Technical Committees set up by Sector Technical Boards (e.g. BTS 1: Building and Civil engineering)
SCs	Sub-committees set up by Technical Committees
WGs	Working Groups, established by TCs and SCs to carry out specific pieces of work
BT	Technical Board, responsible for managing the whole technical programme of CEN.

Theoretically, once a Euro Code or Standard is published in its final form, the equivalent British Code or Standard will cease to function and will be withdrawn. However, in practice it is likely that certain sections of the BS documents will be retained on the grounds that they have special application to the UK.

1.11 HEALTH AND SAFETY REGULATIONS AND PRODUCT SPECIFICATION

I feel it is desirable to draw the attention of readers to the need to observe recognized safety precautions when using certain materials and all types of equipment.

Concrete itself is not a 'hazardous' material. However, Portland cement, when mixed with water is highly alkaline (it has a *pH* of about 13.5), and is considered a caustic alkali which can cause burns to the skin, particularly in people who are vulnerable to dermatitis. A 'safety warning' is included as an Appendix in all British Standards for Portland cement, which states:

> Dry cement in normal use has no harmful effect on dry skin. Precautions should be taken to avoid dry cement entering the eyes, nose and mouth, and to prevent skin contact with wet cement.

Polymer resins are now widely used in the construction industry, particularly in repair mortars, and in coatings, bonding agents and adhesives. There are certain hazards associated with the use of some of these compounds, and users should contact the manufacturers and/or FeRFA (Federation of Resin Formulators and Applicators). This organization is also known as the Trade Federation of Specialist Contractors and Materials Suppliers to the Construction Industry. The 'hazards' associated with the use of these compounds include: contamination of the skin and eyes; harmful effects of inhalation of vapour or mist; fire and explosion. The degree of hazard varies from one compound to another.

Reference should also be made to the Health and Safety Executive and the requirements of this Authority.

The Construction (Design and Management) Regulations 1994 are intended to supplement health and safety legislation and were initiated by the European Union. The first part came into effect in March 1995, and by the end of December 1995, all the remaining provisions became effective.

> The Regulations are proactive, making people assess risks and take precautions, rather than dealing with problems once they have occurred. They target the health and safety of those who build, maintain, install and demolish buildings or plant, and create a structure of health and safety documentation administrated by a planning supervisor and principal contractor.
>
> Construction work is broadly defined in the Regulations and includes:
>
> - demolition, site clearance, construction;
> - maintenance, alterations, renovation;
> - assembly of prefabricated elements to form a structure;
> - cleaning windows, translucent walls and roofs;
> - installation, maintenance, repair and removal of mechanical, electrical and telecommunications equipment including computers [...]
> - preparation, assembly, and removal of any structure including any building [...] bridge, pipeline, sewer, road, waterworks ...[3]

The Construction Products Regulations came into force at the end of 1991, to implement the Construction Products Directive. The potential scope of the Directive is very wide indeed as it is applicable to all types of product that are intended for permanent incorporation in buildings and civil engineering works.

According to *BSI News*[4] the legislative changes prepare the way for the introduction of new forms of:

- product specification:
  - Harmonized European Standards;
  - European Technical Approvals and possibly national technical specifications recognized by the Commission as meeting the essential requirements of the CPD.
- technical information:
  - manufacturer's declaration of conformity;
  - Independent certificate of conformity.

It will provide for the application of the European Community regulatory mark – the CE mark – to construction products.

The test of fitness under the Directive and the UK Regulations involves products having performance characteristics that enable the works in which they are incorporated to satisfy one or more of the six essential health and safety requirements set out in the Directive.

Except when specifically stated, it is not sure when all the above Directives will in fact come into force, but the new arrangements are likely to come into operation gradually over an extended period of time. However, I recommend that all those connected with the use of construction materials follow the progress of these regulations by contacting BSI Quality Assurance and/or the Construction Technology Division of the Department of the Environment.

BSI emphasize that 'the Kite Mark will continue to play an important part in demonstrating a level of quality above the minimum legal requirements'.

Building Research Establishment Information Paper No. IP.11/93 gives information on Ecolabelling of building materials and building products.

1.12 DEFINITIONS

The following Definitions apply to terms used in this book relating to some of the properties of concrete:

Absorption	A process in which one substance penetrates into the body of another, generally by capillary action.
Permeability	The characteristic of a material which allows fluids to pass through it under differential pressure.

Porosity	The pore space in a material as a percentage of the volume of that material.
Surfactant	A material which helps a liquid to flow easily over the surface of a solid.
Viscosity	The resistance of a material to flow.

The following comments are relevant:

While absorption should not be used to determine the quality of concrete, good-quality concrete has a relatively low absorption. BS 5911: part 100 – Precast Concrete Pipes, requires the 24-hour absorption not to exceed 6.5% when tested in accordance with Appendix E of that Standard.

Permeability determines the rate at which water penetrates into and saturates concrete; this has a direct bearing on a number of important characteristics of concrete, such as frost resistance, watertightness, resistance to attack by aggressive chemicals and resistance to penetration by solutions of chlorides.

Reference should also be made to BS 6100: Glossary of Building and Civil Engineering Terms, particularly Part 6 – Concrete and Plaster.

1.13 REFERENCES

1. This outline is taken from the *News Letter of the UK Register of Expert Witnesses*, **2**, Dec., 1995.
2. New Civil Engineer, 23 Feb., 1995.
3. Taylor Joynson Garrett, Solicitors (1995) The Construction (Design and Management) Regulations 1994, *Construction Review*, **1**, 4.
4. *BSI News*, Aug. 1992.

1.14 FURTHER READING

Alternative dispute resolution (ADR):

McKenna & Co. Solicitors, London (1989) *Law Letter*, Autumn/Winter, pp 14, 15 (1995).

McKenna & Co. Solicitors, London (1995) *Litigation Update*, May, Alternative Dispute Resolution, 6–9.

Hollands, D.E. (1992) Alternative Dispute Resolution, *Journal CIArb*, Feb., 57–9.

The expert witness

Newman, P. (1993) Professional liability of expert witnesses, *Journal CIArb.*, Aug. 173–81.

Lord Taylor (1995) The Lund Lecture – The Expert Witness, *Journal CInstArb*, **61**(2), May, 113–17.

EEC construction legislation

Building Research Establishment (1993) Information Paper IP.11/93, Ecolabelling of building materials and building products; May, 4.

Building Research Establishment (1994) Digest 397, Standardisation in support of European legislation: what does it mean for the UK construction industry? Sept., 8.

The Concrete Society (1995) *Seminar, European Standards for Concrete Repair*, 26 April 1995, London, 9 papers.

Taylor Joynson Garrett Solicitors (1995) The Construction (Design and Management) Regulations 1994, *Construction Review*, Issue No. 1, 4.

2

Basic characteristics of concrete and mortar and their constituent and associated materials

2.1 INTRODUCTION

The characteristics of concrete and mortar depend largely on the characteristics of the materials used in their manufacture; the associated materials are needed to enable the concrete and mortar to be used for specific purposes.

Constituent materials

The constituent materials are:

- Portland cements – BS 12 etc and ENV 197–1;
- high alumina cement (HAC) – also known as calcium aluminate cement;
- corrosion-resistant cements;
- aggregates from natural sources for concrete and mortar;
- admixtures;
- water.

Associated materials

The main associated materials are:

- steel reinforcement;
- non-ferrous metals;
- joint fillers and sealants;
- polymers and reactive resins.

The information given in this chapter is intended to be of a general nature and users of the materials listed should always refer to the latest edition of the relevant National Standard.

CONSTITUENT MATERIALS

2.2 PORTLAND CEMENTS (European Standard ENV 197–1)

Portland cement consists mainly of compounds of calcium silicate and calcium aluminate, the calcium silicates are predominate being between 55% and 85%. There is also tricalcium aluminate, 7% to 12% and ferrites 6% to 10%.

It is made by burning at high temperature a mixture of chalk and clay in a rotary kiln. The clinker is ground, and gypsum is added to control the set. BS 12 limits the amount of sulphur (expressed as SO_3) to 3.5%. The fact that Portland cement contains sulphate is important when investigating the possibility of sulphate attack on the concrete or mortar, which is discussed in chapter 3. Further information on the reaction between sulphates of various bases (calcium, magnesium and ammonium) are given later in this section.

The hydration of the cement (the addition of water), results in a complex chemical reaction accompanied by the evolution of heat.

Revised British Standards for cement were published in 1991 and included BS 12: Portland cement and BS 4027: Sulphate-resisting Portland cement.

The new designations for Portland cements likely to be used for repair are as follows:

- Portland cement-class 42.5; to BS 12: 1991 (CEM 1)
- Portland cement-class 52.5; to BS 12: 1991
- Portland cement-class 42.5R; to BS 12: 1991
- Sulphate-resisting Portland cement-class 42.5; to BS 4027: 1991
- Masonry Cement: BS 5224: 1995 – ENV 413-1

The letter R denotes high early strength.

The revisions were mainly concerned with methods of test and terminology and were intended to agree with the European Standard for cement (ENV 197-1). Minor changes in composition were also introduced.

If a cement equivalent to 'ordinary Portland is required, then this should be ordered as 'Portland cement – class 42.5, to BS12: 1991''. If a rapid hardening Portland cement is required, equivalent to 'rapid hardening Portland cement', then a Portland cement-class 52.5 to BS 12: 1991, or Portland cement-class 42.5R to BS 12: 1991 should be ordered.

The above listed cements are the ones used almost exclusively for repair work.

In 1990 and 1991, a completely revised edition of BS 5328: 1991: Concrete, was issued in four Parts.

Blended cements consisting of mixtures of Portland cement and pulverized fuel ash (pfa) and Portland cement and ground granulated blastfurnace slag (ggbs) are used in concrete for special purposes, but I have not come across their use in repair mortars and normal repair concrete.

The principal characteristics of Portland cement are as follows.

1. A very fine powder, particle size 1–50 microns.
2. The paste (cement and water) is highly alkaline, having a pH of about 13.5. This high alkalinity is relevant to the occurrence of alkali-aggregate reaction. The interaction between the alkalis in the cement and certain types of silicious aggregates is discussed later in this chapter under section 2.5, and in chapter 3, section 3.5.13.
3. The setting time (initial and final) is in the range of 45 minutes to 10 hours.
4. Both setting, and hardening (rate of gain of strength) are affected by temperature; an increase in temperature speeds up the chemical reaction between the cement and the mixing water.
5. Portland cement provides a comparatively high compressive strength to concrete and mortar. Tensile strength is only about 10% of the compressive strength.
6. The compounds which are responsible for the cementing action of the cement paste are mainly the calcium silicates (the C2S and the C3S).
7. It is the hydration products of the cement which, other things being equal, determine the strength of the concrete/mortar. The hydration products are very complex chemical compounds, the principal compounds are calcium silicate gel, calcium hydroxide (about 20%) and tricalcium aluminate hydrate. Calcium hydroxide ($Ca(OH)_2$) is liberated by the hydrolysis of the calcium silicates. The various hydration products hydrate at different rates, but the hydration is rapid to start with and then slows down.
8. The two major factors which influence the rate of gain of strength are its chemical composition and its fineness. With modern cements the increase in strength after the first 28 days is likely to be very small and should generally be ignored.
9. The amount of water in the mix (usually referred to as the water/cement ratio) is a vital factor in determining the strength, permeability and absorption of the concrete/mortar. For higher strength and durability the w/c should not exceed 0.50, and for special purposes, 0.40–0.45; this is the free w/c which means the aggregates are saturated but surface dry.

2.2.1 The action of acids on Portland cement

The cement is very vulnerable to attack by acids. The reaction between the acid and the cement takes place immediately the two (acid and hydrated cement) are in contact. While the severity of attack is influenced significantly by the pH of the acid, the chemical composition of the acid is also important. Generally, mineral acids, such as nitric, sulphuric, hydrochloric etc. are more aggressive than organic acids in equal concentrations. See also Chapter 3, section 3.5.2.

2.2.2 Solutions of sulphates and their effect on Portland cement

Reference has been made to the reaction between Portland cement and solutions of sulphates of various bases. Calcium sulphate is only moderately soluble (a saturated solution is formed at about 1100 pm). The formation of gypsum by the reaction between calcium hydroxide and sulphate solutions, more than doubles the volume (Lea, 1970).[1] Lea also stresses that the combination hydrated calcium aluminate and gypsum in solution forms the compound calcium sulpho-aluminate (ettringite) and this also doubles the solid volume. These chemical reactions lead to the expansion and disruption of concrete and mortar.

Magnesium sulphate is much more soluble than calcium sulphate, and has a more destructive action than other sulphates except ammonium sulphate which is probably the most destructive of all sulphates. See also chapter 3, section 3.5.3. Magnesium sulphate reacts with and decomposes the hydrated calcium silicates, as well as reacting with calcium hydroxide and aluminates. See also chapter 3, section 3.5.4.

2.2.3 The effect of solutions of chlorides on Portland cement

Chlorides of calcium, sodium and potassium in normally found concentrations do not attack Portland cement, but they cause corrosion of ferrous metals. However, chlorides react with the tricalcium aluminate hydrate (C3A) in Portland cement to form a compound which tends to inhibit the chlorides from attacking ferrous metals, e.g. steel reinforcement, and this is discussed again in chapter 3, Section 3.5.6.

2.3 HIGH ALUMINA CEMENT (HAC)

HAC is a hydraulic cement based on calcium aluminates rather than calcium silicates which are the main compounds in Portland cement.

The main constituents of HAC, as given by Ciment Fondu Lafarge are:

aluminium oxide (AL_2O_3)	38–40%
calcium oxide (CaO)	37–39%
oxides of iron (Fe_2O_3) etc.	15–18%

It is made by fusing/sintering a mixture of bauxite and limestone.

HAC is a rapid hardening cement, but the setting time is similar to that of Portland cement.

Hydrated HAC paste has a pH of about 11.3 (compared with the pH of Portland cement of about 13.5. The British Standard for HAC is BS 915: Part 2; Metric units: 1972 (1983). The BS requires that the fineness should not be less than 225m^2/kg.

HAC does not release free lime when mixed with water (as does Portland cement) and this results in improved resistance to attack by dilute acids but it is vulnerable to attack by caustic alkalis in solution as the alkalis react with the hydration products of the HAC to form calcium hydroxide and other compounds. If carbon dioxide is present, the calcium hydroxide is converted to calcium carbonate in the presence of moisture, and further complex reactions occur which weaken the concrete and lower the pH. It should be noted that these reactions only take place to a significant degree when the concrete is 'poor' quality (not fully compacted, high w/c ratio, inadequately cured).

Good-quality HAC concrete which is not in contact with caustic alkalis is more resistant to sulphate attack than Portland cement, as shown by the Building Research Establishment long-term trials at Northwick Park Hospital.

Special factors relating to the use of HAC concrete include the following:

1. Water/cement ratio: this must not be less than 0.40; this is critical and is the 'free' w/c, namely the gauging water plus the water on the surface of the aggregate particles.
2. Cement content: the minimum recommended is 400kg/m^3, which is considerably higher than normally used in Portland cement concrete.
3. Admixtures: these should only be used with the advice of the manufacturers.
4. Curing: due to the rapid evolution of the heat of hydration, non-load-bearing formwork could be removed approximately four hours after mixing. Water curing should be carried out so that the surface of the concrete is kept continuously wet for 24 hours after casting.
5. Compaction: as with Portland cement concrete, thorough compaction is essential to obtain maximum strength and low absorption and reduced permeability.
6. Conversion: hydrated HAC undergoes a mineralogical change with the passage of time. This is the conversion of the low-density hydrate system to the high-density stable form, and results in a reduction in compressive strength followed by some recovery of strength. The time taken to reach minimum long-term strength and the relationship between this minimum and the 24-hour maximum depends

mainly on the temperature and humidity conditions under which the concrete operates, and the original water/cement ratio which, as stated above, must not exceed 0.40.

For example, Midgley has shown that concrete cured and stored in outdoor conditions typical of Western Europe is likely to reach its minimum long-term strength in about five years. The minimum long-term strength achieved depends on a number of complex factors. A reasonable figure would be about 50% of its 24-hour design strength. Under certain conditions, conversion can also result in reduced chemical resistance of the concrete.

Due to this very high early strength it was used in the UK almost exclusively for precast prestressed beams. However, in the early 1970s there was a series of failures and the detailed investigations highlighted the serious results of failing to follow recommended practice in the manufacture of the concrete, mainly the use of too high a water/cement ratio. This resulted in the withdrawal of approval in Codes of Practice for the use of HAC concrete for structural elements.

HAC has been successfully used for refractory concrete, floor toppings and slabs to resist acid attack, and special areas subject to abrasion and thermal shock. For these special purposes HAC concrete is often made with Alag aggregate which is based on HAC clinker.

Very rapid setting of HAC can be obtained by mixing it with Portland cement or hydrated lime, but this results in a considerable reduction in the 24-hour strength and should only be used for special purposes not associated with strength or durability.

2.4 CORROSION-RESISTANT CEMENTS

These cements are all proprietary materials and must be used with great care and with special attention to the instructions of the manufacturer and the requirements of the Health and Safety at Work Acts.

There are three types of chemically resistant cements, namely

1. those based on synthetic resins;
2. those based on silicates of sodium and potassium;
3. those based on natural rubber and synthetic rubber latexes.

Synthetic resin cements are resistant to a wide range of chemicals, including acids, alkalies and solvents. The resins used include polyester, phenolic and furane. Mortars based on polyester resins can give exceptionally high compressive strengths (90–100 N/mm^2).

Silicate cements are resistant to a wide range of acids and to very high temperatures, but are generally not resistant to alkalies. These cements are used with a syrup.

Natural and synthetic rubber cements are only moderately resistant to chemical attack and are used when the aggressive conditions are mild. They exhibit low porosity and low permeability and provide a good bond with the substrate. They consist of a prepacked powder and a separate syrup as the gauging liquid. The compressive strength is low, in the range of 8–15 N/mm^2.

For further information on these special cements reference should be made to the manufacturers, Prodorite Ltd, Wednesbury, West Midlands.

2.5 AGGREGATES FROM NATURAL SOURCES FOR CONCRETE AND MORTAR

The relevant British Standard for concreting aggregates is BS 882, and covers gravel and crushed rock and sand.

The British Standard for fine aggregate (sand) for mortars used for repair of concrete is BS 1199: Building Sands from Natural Sources for External Rendering and Internal Plastering. There are two grading tables, one for Type A sands which is coarser than Type B and is the one normally used for mortars for repair work. Sands for floor screeds should comply with BS 882, Table 5, normally grading limit M.

The Standards should be referred to for detailed information on their requirements. Apart from the tables giving grading limits, there are requirements for Flakiness, Limits on Shell Content, and Limits on Clay, Silt and Dust and Chloride Content. BS 882, Appendix B states:

> No simple tests for durability and resistance to frost or wear of concrete can be applied; hence experience of the properties of concrete made with the type of aggregate in question and a knowledge of their source are the only reliable means of assessment.

With sea-dredged aggregates special attention should be paid to the shell and salt (mainly sodium chloride) contents.

In the UK aggregates from some sources in Scotland and the north of England possess high shrinkage characteristics. When there is any doubt about an aggregate, tests should be carried out in accordance with BS 812: Part 120 which details test methods for determining drying shrinkage of mortar prisms made with suspect aggregates and recommendations are given for the interpretation of the results.

I have found differences of opinion on the effect of absorption of aggregates on the permeability of concrete. Code of Practice for water-retaining structures (BS 8007), places a limit of 3% on the absorption of aggregates. I have not found any published information based on properly conducted tests which justify this restriction. In fact the only research report on this subject found, is by Arnold, S.R. and Littleton, I. of the Military College of Science and Technology, Shrivenham, dated

May 1983. This states that a preliminary test programme using concrete with a w/c ratio of 0.5 proved the concrete to be impermeable. The cement contents were in the range 317–30kg/m^3, and the absorption of the aggregates varied from 0.6% to 8%. This justifies the assumption that with a reasonable cement content and a low w/c ratio, the absorption of aggregates in general use has no significant effect on the concrete to resist the passage of water under pressure.

2.6 ADMIXTURES

An admixture can be defined as a chemical compound that is added in very small quantities to concrete, mortar or grout at the time of batching or mixing, to produce some desired characteristic or modification to the mix and/or to the mature concrete, mortar, or grout.

While the use of admixtures in the UK has increased significantly in recent years, this country still lags far behind Continental Europe, the USA and other developed countries. There is an increasing demand for higher quality/performance concrete and properly selected admixtures are more or less essential to achieve this.

The main types of admixtures in general use are:

- accelerators;
- set retarders;
- water-reducers/workability aids/plasticizers;
- superplasticizers;
- air-entraining admixtures;
- pigments.

It should be noted that BS 8110 – the Structural Use of Concrete, refers to pigments as an admixture; but PD.6534: 1993, Guide to the use in the UK of ENV206: 1992 – Concrete, clause 4.5 includes pigments under the heading of 'Additions'. In this book they are included under 'Admixtures'.

The general use of admixtures is covered by various Codes, and by BS 5328, Parts 1 and 3, and by ENV 206 (draft European Standard). The ENV puts an upper limit on the use of admixtures in a mix at 5% by mass of the cement, and a lower limit of 0.2%. The ENV also requires that admixtures in liquid form, when the dosage exceeds 3 litres/m^3 of concrete, shall be taken into account when calculating the water/cement ratio of the mix.

The British Standards are performance specifications.

2.6.1 Accelerators

These are covered by BS 5075: Part 1 and ENV 934–2 (draft European Standard). These only cover accelerators for concrete; there is no BS for accelerators for mortar and grout.

BS 5075: Part 1, defines an accelerating admixture as:

> A material that increases the initial rate of reaction between cement and water & thereby accelerates the setting and early strength development of the concrete.

Some accelerators contain chloride as an active ingredient and when this is the case, the BS requires that the chloride content shall be stated by the manufacturer. This is important because Standards now strictly limit the chloride iron content of concrete which contains ferrous metals (see BS 5328: Parts 2 and 3).

The performance requirements include stiffening times and minimum compressive strength as percentage of the control mix. The 24-hour strength is set as 125% and the 28-day strength as 95%.

2.6.2 Set retarders

A retarding admixture is defined in BS 5075: Part 1 as:

> A material that decreases the initial rate of reaction between cement and water and thereby retards the setting of concrete.

For mortars, a similar definition is applicable. The relevant UK Standard is BS 4887: Part 2. The draft European Standard is ENV 934-2. It should be noted that BS 4887: Part 2, states that the Standard covers building mortars and rendering, but not mortar for floor screeding.

The retarders are in the form of liquids and are organic compounds and the exact composition varies from manufacturer to manufacturer.

2.6.3 Water-reducing admixtures/workability aids/plasticizers

For concrete, these admixtures are covered by BS 5075: Part 1 which defines a 'normal water-reducing admixture' as:

> A material that increases the fluidity of the cement paste without significantly affecting the air content and thereby increases the workability of the concrete at constant water/cement ratio, or permits concrete to be made with a decreased amount of water while maintaining equal workability, with a consequent increase in strength.

The Standard also covers 'accelerating water-reducing admixtures' and 'retarding water-reducing admixtures'. These admixtures combine the two described functions.

At the time of writing this book there did not appear to be an equivalent ENV (draft European Standard) under preparation.

2.6.4 Superplasticizing admixtures

These admixtures are covered by BS 5075: Part 3, and by the draft European Standard ENV 934-2. The UK Standard defines such admixtures as:

> An admixture, that when added to an hydraulic binder concrete, imparts very high workability or allows a large decrease in water content for a given workability.

The increase in workability is dramatic, as the concrete flows, and this is measured as described in BS 1881: Part 105. This super workability only lasts for a limited period, generally about two to four hours.

The very high workability obtained (150–200 mm slump) ensures that the concrete is virtually self-compacting.

Concrete containing superplasticizers is used for a number of purposes which include:

- ultra-high strength concrete;
- placing concrete in locations where compaction is difficult, for example, in members containing congested reinforcement;
- for laying large floor areas in a continuous operation with a comparatively small labour force.

The basic principles underlying the improvement in workability in normal plasticizing admixtures apply to superplasticizers but on a much greater scale. Simply expressed, the admixture particles are negatively charged and are adsorbed onto the surface of the hydrating cement particles, which also become negatively charged. As negatively charged particles repel each other, the cement particles are dispersed and workability thereby is greatly increased.

The two main basic types of superplasticizers are sulphonated naphthalene-formaldehyde condensates, and sulphonated melamine-formaldehyde condensates.

2.6.5 Air entraining admixtures

This type of admixture is defined in BS 5075: Part 2 as:

> An admixture that causes a controlled and stable quantity of air to be incorporated during the mixing of concrete, without significantly effecting the setting of the concrete.

The draft European Standard is ENV 934-2.

The action of the entrained air is to either reduce the water requirement of the mix with constant workability or to increase the workability with a constant water/cement ratio. The presence of the entrained air also increases the resistance of the concrete to freezing and thawing (frost attack). However, there is a reduction in compressive strength compared with a control mix of the same mix proportions and water/cement ratio.

The actual reduction in compressive strength depends on a number of factors, but a figure of 4% for each 1% of air entrained is often used as a guide. It should also be noted that if the cement content of the mix exceeds about 350 kg/m^3 difficulties are likely to arise in entraining the air. The size of the bubbles of entrained air is about 50 microns or 0.05 mm.

Air entraining admixtures for use in mortars are covered by BS 4887: Part 1, in which they are referred to as 'air-entraining (plasticizing) admixtures'. These mortars are used for bedding masonry units and for rendering. In addition to improving workability, the air entrainment also increases the frost resistance of the mortar.

2.6.6 Pigments

The relevant British Standard is BS 1014: Pigments for Portland cement and Portland cement products.

The Standard sets out requirements for sampling, use as a powder or as a dispersion, composition, and the effect of the pigment on the setting time and strength of Portland cement products.

Table 1 of the Standard lists seven pigments, of which four are oxides of iron, one is carbon black, one is chromic oxide and one is titanium dioxide.

The principal pigments in use are oxides of iron and carbon black, and are in the form of very fine powders. The average values for the particle sizes suitable for use in concrete and mortar are:

Iron oxide:	0.1 microns
Carbon black:	0.1 to 0.4 microns

These can be compared with Portland cement – 1.0 to 50 microns, and sand 150 to 5000 microns (1 micron = 0.001 mm).

While pigments are not used to any great extent in the repair of concrete, their use, when necessary, can give rise to complaints and disappointment. This is usually due to lack of understanding of the limitations of pigmented concrete/mortar compared to pigmented coatings. For example, complete uniformity of colour cannot be obtained due to a number of factors, of which the principal are:

1. small but unavoidable variations in the actual mix proportions (cement, aggregates and water);
2. small variations in the colour of the aggregates, particularly the sand, and to a lesser extent the cement;

3. small but unavoidable variations in the actual distribution of the pigment in the mix;
4. variations in the curing regime.

2.7 ADDITIONS

These are materials which are permitted to be added to a concrete or mortar mix in much larger quantities than admixtures. In BS 5328: Part 1, the following materials are included under the heading of Additions in Clause 3.6:

- pulverized fuel ash (pfa) complying with BS 3892;
- ground granulated blastfurnace slag (ggbs) complying with BS 6699.

2.7.1 Pulverized fuel ash (pfa)

Pulverized fuel ash is also classed as a cement replacement and in fact that is its principal use in the concrete industry.

This material is a by-product of pulverized coal-fired electricity generating stations. It is a fine powder, the particle size being in the range 1–50 microns. The approximate composition is :

50% silicon (SiO_2)
28% alumina as $A1_2O_3$
11% iron oxide as Fe_2O_3
11% oxides of calcium, magnesium, sodium and potassium

The relevant British Standard is BS 3892: Parts 1 and 2.

Part 1 covers pfa for use as a cementitious compound in structural concrete; Part 2 covers pfa for use in grouts and for miscellaneous uses in concrete.

Pfa exhibits pozzolanic activity and BS 3892: Part 1 contains a test method for determining the pozzolanic activity index. The Standard limits the sulphate (as SO_3) content to a maximum of 2.5%.

Part 2, which deals with pfa for use in grouts, specifically states that this does not apply to grouts used in ducts for prestressing tendons.

The main advantages claimed for the inclusion of pfa in concrete are:

- reduction in heat of hydration;
- improved workability with constant water/cement ratio;
- increased resistance to sulphate attack;
- reduced permeability to liquids;
- long-term increase in compressive strength.

There is some reason to believe that the inclusion of pfa in the concrete may render it more resistant to alkali–silica reaction.

Reference should also be made to BS 6588: Specification for Portland pulverized fuel ash cement which lays down requirements for composition, strength, physical and chemical properties for two combinations of Portland cement and pfa.

As far as the author is aware, it is not possible to determine by chemical analysis the proportion of pfa in hardened concrete or mortar. The presence of pfa in concrete can be confirmed by microscopic examination of thin sections.

The pfa content in fresh concrete can be determined by the chemical method described in BS 6610, or by the particle density method described in Annex D in Part 128 of BS 1881 – Methods for Analysis of Fresh Concrete.

2.7.2 Ground granulated blastfurnace slag (ggbs)

The relevant British Standard is BS 6699: Specification for ground granulated blastfurnace slag for use with Portland cement. The slag is a waste product from steelworks; the raw materials going into the blastfurnace are iron ore, limestone and coke. The products of the blastfurnace are iron and slag. The slag can be used as an aggregate or as an addition to Portland cement for concrete. When used in combination with OPC it increases the resistance of the concrete to sulphate attack, and to ASR by limiting the alkali content of the binder (cement plus ggbs). The proportions used with OPC depends on the required effect on the resulting concrete. Generally, mixes containing 40% ggbs and 60% OPC to 65% ggbs and 35% OPC are used.

2.7.3 Condensed silica fume

Condensed silica fume is a waste product of the ferrosilicon industry. It consists of 88–98% of silicon dioxide (SiO_2) with very small percentages of carbon, ferric oxide, aluminium oxide (alumina) and oxides of magnesium, potassium and sodium.

It is a very fine greyish powder with a specific surface about fifty times that of normal Portland cement, the particles having a diameter of about 1 micron, and is a highly reactive pozzolan.

The addition of condensed silica fume to concrete and mortar has a significant effect on the properties of the plastic mix, as well as on the hardened material.

The dosage is generally in the range of 2% to 10% by weight of cement.

Its presence imparts a number of beneficial characteristics to the concrete and mortar:

- increased cohesion;
- reduced permeability;
- increase in compressive strength;
- increased resistance to sulphate attack (except possibly ammonium sulphate);
- increased resistance to a number of aggressive chemicals; including some ammonium-based fertilizers.

The very small particle size increases the water demand of the mix and can result in premature stiffening if placing and compaction is delayed. It is normally used with a superplasticizer.

When used with selected Portland cement, selected aggregates and a superplasticizer, compressive strengths of over 90 N/mm^2 can be obtained; this involves careful mix design and strict control of all aspects of the concreting procedure.

In the UK and USA condensed silica fume is marketed as a stabilized slurry which contains a plasticizer or superplasticizer. The Agreement Certificate for the 'Elkem Microsilica' gives the pH of the silica/water slurry as 5.5 plus or minus 1.0.

As far as I am aware, it is not possible to determine by chemical analysis, the proportion of condensed silica fume present in hardened concrete or mortar.

There is no British Standard for condensed silica fume.

2.7.4 Polymers

The term 'polymers' includes a wide range of materials, but in this chapter epoxies, polurethanes and polyesters, are dealt with under section 2.14 – Reactive Resins.

The polymers described in this section are mainly available in liquid or powder forms. The liquids are dispersions (also referred to as latexes), and are generally whitish in colour. The solid content and viscosity vary; the solid content is generally in the range of 40–70%. The polymers available in powder form are mainly:

- modified polyvinyl acetates (PVAs);
- ethylene vinyl acetate (EVAs);
- acrylics.

The dispersions include styrene butadiene rubber (SBR), Neoprene (an artificial rubber), styrene acrylics and natural rubber latexes.

In my opinion, the manufacturers of polymers for use in concrete or mortar mixes should comply with Code requirements for the disclosure of information on ingredients.

While each of the polymers mentioned above have their own specific characteristics, they are all intended to impart some useful properties to the concrete/mortar in which they are incorporated, and these properties are summarized below:

- improved workability of the mix with constant w/c ratio or reduced w/c ratio with constant workability;
- increased bond with the substrate;
- reduced permeability and absorption;
- improved resistance to carbonation;
- some limited increase in resistance to attack by aggressive chemicals.

A great advantage in the use of synthetic polymer dispersions is that they can be tailor-made to meet the requirements of the proposed end use.

The particle size of the SBR solids in an SBR latex is in the range 0.13 to 0.23 microns which can be compared with Portland cement of 1–50 microns.

When these polymer latexes are added to the mix on site, the proportions recommended by the suppliers should in general be followed. In the case of SBRs, the proportions are usually in the range of 5% to 10% by weight of cement; the amount of emulsion added depends on the reason for its addition.

All polymers are expensive and therefore their use should be given careful consideration.

Proprietary repair mortars consisting of prepacked cement and aggregate often have a selected polymer latex included as a gauging liquid to which water must be added to provide the required workability. Other prepacked mixes have the polymer included in powder form.

2.8 WATER FOR MIXING CONCRETE AND MORTAR

The relevant British Standard is BS 3148: Methods of test for water for making concrete (including notes on the suitability of the water).

The water used for mixing concrete and mortar (also known as gauging water), should be free from compounds which injuriously affect the setting and hardening of the mix and/or have an adverse effect on the properties of the hardened concrete/mortar. The impurities may be organic and/or inorganic. The Standard recommends making mortar cubes using the water under investigation with control cubes made with distilled water. Sulphates in solution (as SO_3) should be limited to 1000 ppm. In addition, the maximum concentration of sulphate in the concrete or mortar should not exceed that recommended in BS 5328: Concrete, Part 1 – Clause 4.2.3, namely 4–5% by mass of the cement.

It can be said that water which is fit for drinking (potable water) is also fit for making concrete and mortar. On the other hand, water which is unfit for drinking may be quite suitable for making concrete/mortar for specific purposes. When drinking water is not available for construction work, saline (brackish) water, and even sea water may have to be used. This is discussed in Chapter 3, section 3.5.10.

ASSOCIATED MATERIALS

2.9 STEEL REINFORCEMENT

Steel reinforcement for concrete is covered by the following British Standards:

- BS 4449 Carbon steel bars for the reinforcement of concrete
- BS 4482 Cold reduced steel wire for the reinforcement of concrete
- BS 4483 Steel fabric for the reinforcement of concrete
- BS 4486 Hot rolled and processed, high tensile alloy steel bars for the prestressing of concrete
- BS 5896 High tensile steel wire and strand for the prestressing of concrete
- BS 7295 Fusion bonded epoxy coated carbon steel bars for reinforcement of concrete
- BS 729 Hot dipped galvanized coatings on iron and steel articles (this is not specifically for rebars)
- BS 6744 Austenitic stainless steel bars for the reinforcement of concrete

The coefficient of thermal expansion of plain carbon steel is 12×10^{6}.

2.9.1 Galvanized reinforcement

At the time of writing this book there is no British Standard specifically for galvanized reinforcement, although there are a number of Standards for galvanized steel for various purposes.

The object of galvanizing is to provide protection of the rebars in adverse conditions of exposure. Galvanizing consists of coating the steel with zinc by either dipping the steel into tanks of molten zinc (hot-dip galvanizing) or by electrodeposition from an aqueous solution. For rebars, hot dip galvanizing is used, and this process is covered by BS 729: Specification for hot dipped galvanized coating for iron and steel articles.

I am indebted to the Galvanizers Association for the information which follows.

Galvanized reinforcement was first used in Bermuda in the 1930s and then became widely used in the 1940s when sea-dredged aggregates were used for structures during the war years. In the UK it appears to be mainly used in precast concrete units for large building projects.

When concrete/mortar is placed around galvanized rebars there is a chemical reaction between the zinc coating and calcium hydroxide in the hydrating cement paste. The zinc surface is passivated with the evolution of hydrogen. The passivation occurs with the initial formation of a layer of zinc hydroxide; further chemical reactions follow resulting in the formation of a complex stable zinc compound, zincate.

For durability in aggressive conditions, it is essential for this passivated film on the surface of the zinc to be undamaged.

The presence of very small concentration of chromate (about 0.002% or 20 ppm) in the cement will inhibit the reaction between the cement paste and the zinc and thus limit the formation of hydrogen.

The protection of the steel provided by the zinc coating is mainly dependent on the thickness of the coating and therefore the thickness should be specified to meet anticipate exposure conditions.

2.9.2 Fusion-bonded epoxy-coated reinforcement

This method of protecting reinforcement from corrosion in aggressive environments has been in use, mainly in the US, since the early 1970s, while in the UK its acceptance has been much slower.

The American Standard is ASTM A775: Standard Specification for Epoxy–Coated Reinforcing Bars; The UK Specification is BS 7295: Parts 1 and 2: Fusion-bonded Epoxy-coated Carbon Steel Bars for the Reinforcement of Concrete.

The coating is an epoxy powder specially formulated to resist impact and abrasion, and to possess a sufficient degree of flexibility to bending stresses in the detailing of the bars, and to possess high bond to the surface of the rebars. The epoxy resin used should comply with Part 2 of BS 7295 and is defined as a thermosetting epoxy powder coating material consisting mainly of epoxy resin plus curing agent and pigments. Two conditions are paramount for the durability of the steel under corrosive attack (e.g. by chlorides); these are as follows.

1. High bond strength to the surface of the rebars which must be properly prepared to receive the coating. ASTM A775 requires the bond to be tested by bending coated bars through 120° without cracking or debonding.
2. Minimum amount of damage to the coating which requires a minimum coating thickness. ASTM A775 calls for a thickness in the

range of 0.13 to 0.30 mm (130–300 microns). The British Standard makes a similar recommendation for individual measurements.

It is acknowledged that some damage will occur to the coating during transportation, and during fixing, placing and compacting the concrete around the bars. Research has shown that high and consistent bond strength limits significantly the extent and rate of corrosion of the rebars arising from damage to the coating.

2.9.3 Stainless-steel reinforcement

The relevant British Standard is BS 6744: Specification for Austenitic Stainless Steel for the Reinforcement of Concrete. Of the three basic groups of stainless steel – martensitic, ferritic and austenitic – the austenitic steels (types 302, 315 and warm-worked 316) are by far the most resistant to attack by concentrates of chlorides. A BRE publication in 1988 showed that concrete test specimens containing 3% chloride by weight of cement showed no sign of corrosion over a period of 17 years.

The type 316 steel contains 18% chromium, 10% nickel and 3% molybdenum. In continuously warm and humid conditions, such as exist in swimming pool halls, and if the steel is very highly stressed, serious corrosion can occur; this is illustrated by the failure of stainless-steel hangers supporting a reinforced concrete ceiling slab in a swimming pool in Switzerland in 1985.

It can be welded, but is non-magnetic and therefore it is difficult or impossible to detect it with the normal type cover meter. Galvanic corrosion can occur if the stainless steel is in contact with mild steel in the presence of moisture. It is the mild steel which corrodes, i.e. it is anodic to stainless steel.

Stainless steel is much more expensive than ordinary carbon steel and its use for reinforcement is only justified in special cases.

The coefficient of thermal expansion of austenitic steels is $18 \times 10^{-6}/°C$, compared with carbon steel: $12 \times 10^{-6}/°C$.

2.9.4 Spacers

The use of the correct spacers is an essential part of the construction of reinforced concrete. There is no British Standard for spacers, but in 1991 the Concrete Society published a manual, *Spacers for Reinforced Concrete*. Spacers are used to help ensure correct cover to the rebars. A reduced cover (with other things being equal) can seriously reduce the durability of the reinforcement and thus the 'useful life' of the concrete units.

Spacers should meet the recommendations in the manual for load, deflexion and stability; they must be securely fixed in the correct locations. Spacers are normally made of plastics but some are made of fibre-reinforced cement-based material. I favour a material which bonds to the surrounding concrete.

2.9.5 Corrosion inhibitors

This is a relatively new range of materials which has been used in the US and in the EC countries for some years, but is being accepted only slowly in the UK.

There is no British Standard for this type of material and so reference is made to the US Standard, ASTM C-494 Type C (Admixtures for Concrete).

The following is a description of a corrosion inhibitor made by Grace Construction Products, namely DCI-S.

It is a liquid added to concrete during batching. It chemically inhibits the corrosive action of chlorides on steel reinforcement and prestressing tendons in concrete. It complies with ASTM C-494 Type C, and holds an Agreement Certificate No. 96/3232; it is based on calcium nitrite. By chemically reacting with the embedded steel the passivating layer on the steel is maintained unbroken when chlorides are in contact with the steel.

The dosage should be related to the anticipated chloride ion content of the concrete during the design life of the structure. It is compatible with all types of Portland cement and with other standard types of admixtures, which must be added separately. The dosage rate varies from 10 to 30 litres per m^3 of concrete. The concrete should be good quality, minimum cement content of 350 kg/m^3 and maximum w/c of 0.45 and be thoroughly compacted and cured.

I am indebted to W.C. Grace Ltd for the above information.

2.10 NON-FERROUS METALS IN CONCRETE

Only a limited number of non-ferrous metals are likely to be needed in repair work to reinforced concrete structures. When they are required it is usually as fixings in the concrete or as fixtures in contact with the surface of the concrete. When dissimilar metals are in contact in a moist environment there is danger of galvanic action resulting in the corrosion of one of the metals. This is a complex subject and is dealt with in some detail in a BSI publication, PD. 6484: *Commentary on Corrosion at Bimetallic Contacts and its Alleviation*. For example if carbon steel reinforcement is in contact with stainless steel the carbon steel may suffer moderate corrosion.

2.10.1 Aluminium and aluminium alloys

If unprotected by anodizing or suitable coatings, aluminium in direct contact with damp concrete is liable to be attacked by the caustic alkalis in the cement paste. Detailed recommendations for the protection of aluminium against corrosion are given in BS 4873: Aluminium Alloy Windows.

The coefficient of thermal expansion of aluminium is about $25 \times 10^{-6}/°C$.

2.10.2 Copper

Copper is resistant to most conditions met with in building construction and structures for water supply and sewerage works. It was used for many years as water bars in joints in concrete liquid retaining structures but has been replaced by plastics such as polyvinylchloride compounds.

Copper is not attacked by Portland cement concrete unless chlorides are present in higher concentrations than is now recommended by BS 5328: Part 1; Table 4.

It is attacked by ammonium compounds.

If copper is in direct contact with carbon steel in the presence of moisture galvanic action may occur causing corrosion of the mild steel (the steel is anodic to the copper). The main use now for copper is in roofing and copings.

The coefficient of thermal expansion of copper is about $17 \times 10^{-6}/°C$.

2.10.3 Phosphor-bronze

Phosphor-bronze is an alloy of copper and tin with phosphorous as copper phosphide. It is used mostly for fixtures and fittings in conditions which would result in the corrosion of ferrous metals (except stainless steel).

Phosphor-bronze in contact with mild steel and stainless steel in the presence of moisture will not result in galvanic action.

The coefficient of thermal expansion of phosphor-bronze is about $20 \times 10^{-6}/°C$.

2.10.4 Brass

Brass is mainly an alloy of copper and zinc, but other elements such as aluminium, manganese, nickel, tin and lead, are frequently added in small quantities. There may be slight initial attack by the plastic concrete which results in a protective coating on the metal. Further attack is unlikely unless chlorides are present (see notes on Copper, section 2.10.2).

2.10.5 Lead

Lead is vulnerable to attack by Portland cement mortar and concrete and by lime. There is an initial attack from the plastic mortar/concrete, but this will cease after the concrete or mortar has hardened. It may recommence if the concrete/mortar becomes wet. Protection can be provided by the application of a thick coat of bitumen.

2.10.6 Zinc

Zinc reacts with caustic alkalis (such as hydrating Portland cement paste), with evolution of hydrogen, which can result in the spalling of the concrete/mortar, but only when there is a relatively large volume of zinc being attacked. The zinc coating on steel (galvanizing), would not result in any damage to the surrounding concrete/mortar. Zinc rods embedded in concrete or mortar may be damaged, but can be protected by a thick coat of bitumen.

2.11 Joint fillers and sealants

Defects in joints in concrete structures, both insitu and precast, are a major source of complaints arising from unsatisfactory performance. It is therefore essential to locate and design joints correctly and to select the most suitable materials for use as joint fillers and sealants.

Two relevant British Standards are:

1. BS 6093: Code of Practice for design of joints and jointing in building construction;
2. BS 6213: Guide to selection of constructional sealants.

In addition, there are five British Standards covering specific types of sealant and these are referred to later in this section.

For movement joints in the floors of commercial and industrial buildings and in bridges, special proprietary joint assemblies are to be preferred to what may be termed 'in-house' designs. These assemblies generally consist of sides and base of aluminium or brass securely fixed into the structural slab on each side of the joint, with a high-quality synthetic rubber insert to effectively seal the joint at the top surface.

2.12 JOINT FILLERS

These materials are used in 'in-house' designed joints, and are sometimes referred to as back-up materials. They provide support to the sealant but should not bond to it. The joint filler also prevents the ingress into joints in floor and roof slabs of stones and debris during the construction period as the sealant is usually applied later in the contract.

The material used for joint fillers should fulfil the following requirements.

1. It must be durable under service conditions; **ideally**, the life should be the same as the service life of the structure of which it forms part.
2. It must be chemically inert and non-toxic.
3. It must be resilient, but should not extrude so as to interfere with the integrity of the sealant.
4. It should not bond to the sealant; if it does, a bond breaker must be used.
5. It must be easily formed to the correct dimensions and be readily inserted into the joint.
6. Additional properties may be required for special purposes (such as tanks holding potable water, sewage tanks, tanks holding chemical waste, and petroleum oil).

The main materials used for joint fillers are:

1. cork granules bonded in a resin which is suitable for long-term immersion in water;
2. cork granules bonded in bitumen which is suitable for immersion in water, but the bitumen may render it unsuitable for use in tanks holding potable water;
3. wood fibre with bitumen; this type is not suitable for use in damp conditions.

2.13 SEALANTS

These materials can be divided into two basic groups:

1. in situ compounds;
2. preformed compounds.

Materials used in both groups should possess the following characteristics:

1. For external use and in liquid retaining or excluding structures the sealant itself should be virtually impermeable.
2. Ideally, the service life should be the same as that of the structure in which it is used.
3. It must bond well to the sides of the joint in which it is inserted, under site construction conditions, and be relatively easy to install.

4. Special requirements may be required for potable water tanks (such as non-toxic, non-tainting, and should not support growth of bacteria, fungus, algae etc.) For tanks holding chemical waste, petroleum oil, and sewage tanks, other requirements will be needed to ensure durability.
5. It should deform in response to movements in the structure without extruding and without losing its integrity.

British Standard BS 6213 gives guidance in the selection of constructional sealants, but deals almost entirely with in situ sealants and is referred to again in the next sub-section.

2.13.1 In situ compounds

The in situ sealants can be divided into two main classes:

1. hot applied sealants;
2. cold applied (pouring grade and gun-grade).

The hot applied sealants are mainly used for joints in external concrete pavements and horizontal ground floor slabs of industrial and commercial buildings. They are normally based on bitumen.

The cold applied pouring grade sealants are used in various types of structures but are confined to horizontal floor and roof slabs.

The gun-grade sealants are suitable for use in a wide range of locations and structures.

The relevant British Standards for both types of sealants are:

BS 2499	Hot-applied joint sealants for concrete pavements
BS 4254	Two-part polysulphide sealants for the building industry
BS 5212	Cold-poured joint sealants for concrete pavements
BS 5215	One-part gun-grade polysulphide based sealants
BS 5889	Silicone based building sealants

BS 6213, Table 2 lists the 'expected service life' of various types of sealants; one- and two-part polysulphides and silicones have predicted 'lives' up to 20 years, with a note that 'under favourable conditions' this may be exceeded. It is not clear from the Standard what constitutes 'favourable' conditions. My experience is that some polysulphides can give disappointing long-term results. However, BS 6213, in Tables 4–11, does give very useful information on the suitability of some 19 types of sealant for a wide variety of uses.

BS 6093: Design of joints and jointing in building construction gives 'expected' lives rather more optimistically than BS 6213, e.g. polysulphides (gun-grade) up to 25 years; silicones up to 25 years and polyurethane sealants, chemically curing, up to 20 years; hot-poured bitumen/rubber sealants up to 10 years.

2.13.2 Preformed sealants

My experience is that preformed sealants suffer from one serious practical disadvantage, namely, the sides of joints have to be smooth and even as the preformed material does not accommodate itself well to an uneven or out of true surface. It is necessary to use a primer on the sides of the joint prior to the insertion of the preformed sealant.

However, Neoprene and EPDM (ethylene-propylene-Diene-Monomer) are particularly resistant to a wide range of chemicals and to deterioration (degradation) by bacteria and fungus growths, and are particularly suitable for water and sewage tanks.

2.14 REACTIVE RESINS

2.14.1 Introduction

These materials are mainly used for coatings and for thin bonded repairs when mixed with specially graded and selected fine aggregate. The information which follows refers to resins used for the repair of damaged concrete and for protective coatings (barrier systems) against chemical attack.

Apart from the obvious need for the resin-based mortar to be adequately durable for the purpose of its intended use, it is essential that it should bond well to the concrete substrate to which it is applied. The resins listed below all possess excellent bond characteristics, but it must be remembered that the bond at the interface of the two materials (mortar and substrate), depends also on the strength of the concrete. The repair of a relatively poor quality concrete presents serious difficulties and to use an expensive resin based mortar may be a waste of money.

The cost of resin-based mortars is many times that of cement-based material and therefore their use is only justified in special cases. However, this comment does not apply to their use as coatings. Coatings, also known as barrier systems, are dealt with in some detail in Chapter 7.

The great advantage of synthetic resins is that they can be formulated to meet a wide range of special requirements.

The main resins used for mortars for repair of concrete are:

- epoxies;
- polyesters;
- polyurethanes

2.14.2 Epoxy resins

These resins are by-products of the petro-chemical industry, and have been known for a long time before methods were developed to make practical use of them by Casten in Switzerland and Greenlee in the USA.

The basic resin is a liquid with a fairly high viscosity and will remain in this condition almost indefinitely. For practical use it must be mixed with a hardener (sometimes referred to as an accelerator). The hardener reacts chemically with the epoxy and changes it from a liquid to a solid; adequate mixing by mechanical means is essential for effective dispersion of the hardener in the epoxy. Shortcomings in the mixing will result in weak spots in the molecular chains which can lead to failure of the cured compound.

There are many types of epoxy resins and experienced formulators can produce resins with a number of desirable qualities.

The principal properties of epoxy resins are:

- high-bond strength (adhesion) to many materials;
- very low shrinkage during curing;
- good dimensional stability after curing;
- high resistance to a wide range of chemicals;
- high resistance to abrasion when used with selected aggregates;
- high resistance to water penetration;
- high compressive, tensile and flexural strength compared with Portland cement mortar and concrete;
- the coefficient of thermal expansion of a sand-filled epoxy mortar is about two to three times that of concrete made with natural aggregates.
- epoxies suffer a considerable loss of compressive strength with increase in temperature. The resin component of a resin-based mortar can be virtually destroyed by a serious building fire; see also Chapter 6, section 6.13.1. The resin can be formulated to bond well to damp surfaces. A primer should be used.

2.14.3 Polyester resins

Polyester resins are similar to expoxies in many, but not all respects. The main differences are:

- *The coefficient of thermal expansion is about 1.5 times that of epoxies.* The coefficient of thermal expansion of unfilled resin is about 6 to 8 times that of concrete made with natural aggregates. With sand-filled resin mortar, the coefficient is about four times that of concrete.
- Shrinkage during curing is appreciably higher than that of epoxies.
- The shelf life of polyester resins is strictly limited while that of epoxies is almost unlimited. For the resin to harden, a catalyst must be added and this results in the development of strong molecular chains.

- The amount of catalyst used does not require the same accuracy of measurement as that needed for epoxies.

The curing of polyester mortars is adversely affected by the presence of moisture. A primer should always be used.

2.14.4 Polyurethane resins

These are used mainly as the basic resin for floor sealants and coatings. Their use as joint sealants has already been mentioned earlier in this chapter.

There is a large variety of polyurethane resins all of which possess specific qualities. It is therefore of special importance that the performance requirements of the finished product be clearly stated.

Polyurethanes have a number of excellent characteristics including flexibility, resistance to chemical attack, special resistance to abrasion and impact. They bond well to concrete and mortar, but the surface must be dry.

2.14.5 Polymerized concrete

This is quite a different material to polymer concrete, the latter being concrete with the cement replaced by a selected polymer, such as an epoxy resin.

Polymerized concrete can be divided into two types:

1. The complete unit of hardened concrete is immersed in a monomer, removed and then polymerized by heat.
2. The monomer is mixed with the gauging water, and then after the concrete has hardened, the unit is polymerized by heat.

The material is claimed to have very high compressive and tensile strength compared with normal high-quality Portland cement concrete, and to be resistant to attack by a wide range of chemicals and to freeze–thaw conditions.

The process of producing polymerized concrete is expensive and complicated and as far as I have been able to ascertain the material is more of academic than practical interest.

2.15 CURING COMPOUNDS FOR CONCRETE AND MORTAR

The efficient curing of concrete is essential in order to ensure strength, resistance to shrinkage cracking and resistance to abrasion. Materials for curing are in the form of:

1. spray-applied membranes;
2. sheet materials.

2.15.1 Spray-applied membranes

The relevant British Standard for testing spray-applied membranes is BS 7542: Methods of Test for Curing Compounds for Concrete. The US Standard is ASTM C309-81: Standard Specification for Liquid-membrane Forming Compounds.

These compounds are generally either water-based or resin-based. They are applied to the newly laid concrete as soon as the concrete is free from surface water (bleed-water). The concrete surface must be protected until the membrane has dried and care exercised to ensure that the membrane is not damaged for at least four days after application.

Many water-based and resin-based compounds gradually weather off, but if it is intended to apply a coating or other finish for the protection or decoration of the concrete, then the suppliers of the membrane should be consulted as some curing compounds will interfere with the bond between applied finish and the base concrete. The same comment applies to the application of rendering, and bedding mortar and proprietary adhesives for tiles and mosaics. As bond failure of an applied finish would be a very serious matter, I recommend that a trial area be arranged at the start of the work.

Some proprietary curing compounds based on polymer resins are claimed to improve the abrasion resistance of the concrete and reduce the penetration of carbon dioxide and thus slow down carbonation of the concrete. For information on the carbonation of concrete see Chapter 3, section 3.2.5

2.15.2 Sheet materials

The principal sheet material used in the curing of concrete and mortar is polyethylene sheeting (trade name polythene). This is very effective in reducing moisture loss provided it is laid as soon as practicable after completion of the area to be cured, and is securely held down around the edges and kept in position for at least four days. I recommend that at least 1000 gauge sheeting should be used as anything thinner is too easily damaged under normal site conditions.

2.15.3 Wet/water curing

The curing of concrete by water spray is only carried out in special cases where it is desirable to keep the concrete temperature under control, e.g. high alumina cement concrete.

2.16 REFERENCE

1. Lea, F.M. (1970) *The Chemistry of Cement and Concrete*, 3rd edn, Edward Arnold Ltd, London, 725.

2.17 FURTHER READING

Portland cements

Lea, F.M. (1970) *The Chemistry of Cement and Concrete*, 3rd edn, Edward Arnold Ltd, London, 725.

Building Research Establishment (1986) *Changes in Portland Cement Properties and their Effects on Concrete*; IP **3**/86, 3.

Spooner, D.C. (1995) *A Guide to the Properties and Selection of Cements Conforming to British and European Standards*, British Cement Assoc. Tech. Note 13, May, p. 29.

High alumina cement

Building Research Establishment (1994) Assessment of existing high alumina cement concrete construction in the UK, *Digest* 392, March, 12.

Lafarge Aluminous Cement Co. Ltd. *Ciment Fondu Lafarge: Guide for Use*, Ref. CA4 – 890, 8.

Concrete Society (1997) *Calcium Aluminate Cements in Construction*, Technical Report **46**, 64.

Corrosion resistant cements

Prodorite Ltd. (1991) *Corrosion Resisting Cements*, p. 15.

Aggregates

Building Research establishment (1987) *Sea-dredged Aggregates in Concrete*, IP, **7**/87, p. 4.

British Standards Institution, *Testing Aggregates*, BS 812.

Midgley, H.G. (1958) The staining of concrete by pyrite, *Mag. of Concrete Research*, Aug. 75–8.

Building Research Establishment (1991) Deterioration of Mundic concrete houses, *BRE News*, April.

Alkali-aggregate/silica reaction

Building Research Establishment (1991) Alkali aggregate reactions in concrete, *Digest* 330: 8.

British Cement Association (1992) Up-to-date information on AAR, *BCA Bulletin*, July, 6–7.

Hobbs, D. (1992) Deleterious alkali–silica reactivity of a number of UK aggregates, *Concrete*, May/June, 64–70.

Admixtures

Buekett, J. (1994) Admixtures, the UK scene, *Concrete*, May/June, 15–17.
Egan, P.J. (1994) Benefits of superplasticising admixtures, *Concrete*, May/June, 18–21.
Concrete Society (1980) *Guide to Chemical Admixtures for Concrete*, London, 16.

Additions

Building Research Establishment (1991) *Durability of pfa Concrete Structures*, IP, **11**/91, 4.
Building Research Establishment (1992) *Durability of Blastfurnace Slag Cement Concrete*, IP, **6**/92, 4.
French, W.J. (1992) Determination of the ratio of pfa to Portland cement in hardened concrete, *Concrete*, May/June, 43–5.
Malhotra, V.M. and Carette, G.G. (1983) Silica fume concrete properties, applications and limitations, *Concrete International*, May, 40–6.

Polymers

American Concrete Institute (1981) *Applications of Polymer Concrete*, Ref. SP–69, 228.
Concrete Society (1994) *Polymers in Concrete: Tech. Report* **39**, 31.
Dennis, R. (1992) *Polymer Dispersions Construction Materials Reference Book*, ed. D.K. Doran, Butterworth Heinemann.
Plum, D.K. (1990) The behaviour of polymer materials in concrete repair & factors influencing selection, *The Structural Engineer*, **68** (17/4) Sept. 337–45.
Walters, D.G. (1990) Comparison of latex-modified Portland cement mortars; *ACI Materials Journal*, July/Aug., 372–7.

Pigments

Spence, F. (1988) Coloured concrete, *Concrete Forum*, April, 15–18.
Hof, H., Jungk, A.E. and Queckborner, J. (1984) Farbige Betone (Coloured concrete), *Betonwerk + Fertigteil – Technik*, no. 50, Heft 7, 459–67.
American Society for Testing Materials (1982) *Standard Specification for Pigments for Integrally Coloured Concrete*, C979-82.
ASTM (1981) Task Group Sub-committee Section C09.03.08.05, Pigments for Integrally Coloured Concrete, pp. 74-7.

Steel reinforcement

ASTM (1984) *Standard Specification for Epoxy-coated Reinforcement Steel Bars*, ASTM/A775M-84.
Concrete Society (1985) *Spacers for Reinforced Concrete*, ref. CS. 101, p. 30.
Concrete Society (1985) *Steel for Prestressed Concrete*, ref CS 015, p. 8.
Galvanizers Assoc. (1994) *Galvanized Reinforcement*, p. 8.
George Clark (Sheffield) Ltd (1989) *Stainless Steel Reinforcing Bar*, Jan. 1, 11.
Hartley, J. (1994) Improving the performance of fusion-bonded epoxy-coated reinforcement, *Concrete*, Jan./Feb., 12–15.

Joints and joint sealants

American Concrete Institute (1990) *Guide to Seals in Joints for Concrete Structures*, Ref. 504R.

Beech, J.C. (1990) *Joint Sealants and Primers: Further Studies of Performance with Porous Surfaces*, BRE Information Paper IP. 4/90, May, 4.

Construction Industry Research and Information Assoc. (CIRIA) (1987) *Civil Engineering Sealants in Wet Conditions*, Tech. Note 128 p. 76.

CIRIA & Sealant Manufacturers Conference (1976) *Manual of Good Practice in Sealant Application*, 72.

Resins

Vipulanandan, C. and Paul, E. (1990) Performance of epoxy and polyester polymer concrete, *ACI Materials Journal*, May/June, 241–51.

Rizzo, E.M. and Sobelman, M.B. (1989) Selection criteria for concrete repair materials, *Concrete International*, Sept. 46–9.

Tabor, L. (1983) Concrete repair, the role of resin compositions, *Concrete*, Feb., 19 and 22.

3

Factors affecting the durability of reinforced concrete

3.1 INTRODUCTION

When one considers the need for a structure to be 'durable' there immediately arises the questions of what maintenance is required, the time lapse between construction and the need for repairs, and what is the likely 'life' of the structure before partial or complete replacement is needed.

Two relevant British Standards are:

BS 8210 Guide to Building Maintenance Management
BS 7543 Guide to the Durability of Buildings, Building Elements, Products and Components.

Reference can also be made to:

BRE Digest 366; Oct. 1991: Structural appraisal of existing buildings for change of use.

The following terms are important and the definitions given here apply to the statements made in this book.

- **Durability** A material can be considered as durable if it fulfills its intended duty for the whole of its design life with an acceptable amount of maintenance including general repair.
- **Design life** This is the length of time which the designer estimates the material will remain durable.
- **Service life** This is the **actual** length of time the material remains durable.
- **Maintenance** A set of measures which when applied to a structure enables the structure to fulfil its functions during its working life. These measures should include cleaning, minor repairs, repainting/recoating, and replacing parts when required.

Table 1 in BS 7543 sets out suggestions for the design life of various types of buildings, and Table 2 gives design life for various categories of building components.

These suggestions are of course all linked to the anticipated level of maintenance and repair applicable to the type of building and component. It must also be remembered that change of use can result in major structural changes, extensive refurbishment and in extreme cases in demolition and rebuilding.

The causes of deterioration of reinforced concrete structures over a practical range of use and environment are discussed in this chapter. After the causes of deterioration have been established consideration has to be given to repair methods and these must be related to the design life of the repaired structure.

At the present time there is no known construction material which is completely inert to chemical attack and immune to physical deterioration.

While it is necessary to distinguish between the causes of the deterioration of the concrete and of the steel reinforcement, in a reinforced concrete structure these two have to be considered together.

It has been stated in Chapter 1 that 'corrosion of steel reinforcement is the most serious durability problem affecting concrete structures'. This clearly indicates that the steel reinforcement is the 'Achilles heel' of reinforced concrete and this certainly agrees with my experience.

3.2 CORROSION OF STEEL REINFORCEMENT IN CONCRETE

3.2.1 Introduction

The corrosion products of steel (generally known as rust) consist of oxides of iron. For rusting to occur moisture and oxygen must be present, and with steel embedded in concrete at or above ground level, both are present.

The action of acids on steel reinforcement is not considered in detail here, but if acids reach the rebars through cracks or defects in the concrete cover, severe corrosion will occur. Some chemical compounds in use in water treatment plants are acidic in solution, e.g. ferrous sulphate and sodium hydrogen sulphate.

Normal design and construction of reinforced concrete ensures that the reinforcement is not subjected to physical deterioration and it is electro-chemical reactions on the surface of the rebars which result in corrosion/rusting of the steel.

Steel does not corrode when it is surrounded with concrete or mortar which has a pH of not lower than about 12.5. The high alkalinity 'passivates' the steel due to the formation of a protective film of oxidation products, such as ferric oxide.

The integrity of the film is maintained by the continued high alkalinity of the surrounding cement paste and as long as this state exists further oxidation is inhibited and the steel does not corrode.

It therefore follows that subsequent oxidation (corrosion) must be due to the breakdown of the passivation of the steel surface. This loss of integrity of the oxide film can be caused by a number of factors, the principal ones being set out below:

1. physical damage to the concrete surrounding the steel caused by impact or abrasion/wear;
2. development of cracks in the concrete extending down to the steel, caused by shrinkage and/or stress;
3. high permeability and/or high porosity of the concrete surrounding the steel;
4. inadequate thickness of the cover coat of concrete or mortar to resist the ingress of aggressive agents for the 'required' period (generally the design life of the structure or component); Codes of Practice lay down nominal figures for depth of cover;
5. the presence in the concrete or chlorides in excess of the recommended 'safe' concentration, as laid down in Standards and Codes of Practice.

Conditions (1) to (4) above can allow the ingress into the concrete down to the rebars of moisture and chemicals. Carbon dioxide in the air causes carbonation of the concrete which reduces the pH of the concrete to about 9.5.

Additional information is given below on the above items except (1) physical damage, as I feel this does not require special discussion.

3.2.2 Development of cracks in concrete

All concrete contains micro-cracks which do not adversely affect the satisfactory performance of the concrete. However, macro-cracks, due to shrinkage and excessive stress, if they extend down to the steel can lead to loss of passivation leading to corrosion of the reinforcement.

Shrinkage cracks can be due to a high water/cement ratio or the use of shrinkable aggregates (see comments in Chapter 2, section 2.5), and/or inadequate curing of the concrete. There are differences of opinion regarding the maximum water/cement ratio which should be specified. I believe that for any type of concrete containing steel reinforcement the w/c ratio should not exceed 0.5, and for high-quality concrete the w/c should be 0.4. For certain precast products, e.g. concrete pipes, an even lower water/cement is often used, down to 0.35 in some cases. Poorly graded aggregates (too high a proportion of fine material) is likely to result in an increase in water demand for the mix and increased drying shrinkage.

3.2.3 High permeability and/or high porosity

For low permeability, the capillary pores in the concrete should become discontinuous within a period not exceeding 14–28 days after the start of hydration.

If the w/c ratios recommended above are too low to provide adequate workability for full compaction of the concrete, then plasticizers should be used as maximum compaction is essential for low permeability.

Full compaction of the concrete around the rebars is essential for the protection of the steel, by reducing permeability and porosity and for development of strength.

Proper curing of the concrete is also essential, as this reduces the risk of drying shrinkage cracking and promotes hydration of the cement.

Some references are given in the 'Further reading' section at the end of this chapter.

3.2.4 Cover coat of concrete or mortar

It is obvious that for the adequate protection of the rebars the cover to the rebars must have a minimum thickness which will depend on the quality of the concrete and the conditions of exposure. This 'minimum' thickness is usually expressed in terms of 'nominal' cover which allows for practical tolerances for fixing the rebars, and space for placing and compacting the concrete. Nominal cover to rebars is laid down in Codes of Practice and is related to conditions of exposure, quality of the concrete in terms of minimum cement content and maximum water/cement ratio. Permissible tolerances are laid down in the relevant Codes.

3.2.5 Carbonation of concrete

Carbon dioxide in the air reacts with calcium hydroxide in the hydrating cement paste to form calcium carbonate:

$$Ca(OH)_2 + CO_2 = CaCO_3 + H_2O$$

This reaction results in a significant reduction in the pH of the concrete (from about 13.0 to 9.5), and is known as 'carbonation'. When the concrete cover to the rebars is carbonated the high alkalinity of the cement paste (which is needed to preserve passivation of the steel) is thus destroyed leaving the steel vulnerable to corrosion. However, for corrosion to occur, oxygen and moisture must be present as well as the formation of a corrosion cell, i.e. an anode and a cathode, on the surface of the steel. An electric current flows from anode to cathode via the electrolyte which is the moisture in the hydrating cement paste (i.e. the water in the capillary pores). The formation of this cell is caused by the differences in potential along the bar. The corrosion of steel in concrete is thus an electro-chemical phenomenon.

The surface of concrete exposed to the air carbonates almost immediately forming a carbonated layer of micron thickness. Concrete has a pore structure and is classed as 'porous', and this is why liquids and gases can slowly penetrate into concrete. The carbon dioxide thus penetrates into the concrete, the rate of penetration depending on a number of factors, the principal ones being the porosity and permeability and moisture content of the concrete. The formation of this carbonated concrete slows down the rate of penetration of carbon dioxide as the calcium carbonate acts as a pore filler in the hydrating cement paste. Carbonation of concrete does not reduce its strength. The speed at which the corrosion cells are formed depends on the availability of moisture and oxygen.

A great deal has been written on the subject of the carbonation of concrete and its effect on the rusting of the rebars; some references are given in the 'Further reading' section at the end of this chapter. Reference should also be made to Chapter 7 which discusses coatings and barrier systems.

3.2.6 Chloride-induced corrosion of reinforcement

Corrosion of rebars can occur in un-carbonated concrete due to the presence of chlorides, and this is known as chloride induced corrosion. Chlorides are present either because they are added to the concrete mix, or are present in the aggregates and/or mixing water, or penetrate into the hardened concrete from an outside source.

They are then present in solution in the pore water. When a salt is dissolved in water, it is immediately split up into electrically charged particles known as ions: $NaCl = Na^+ + Cl^-$. It is the negatively charged chloride ions which destroy the passivity (the layer of ferric oxide) on the surface of the rebars.

In practice the chloride ions present in the concrete exist in two forms, free chloride ions and combined chloride ions.The combined ions are combined with the hydration products in the cement paste, mainly the tricalcium aluminate (C3A). It is generally agreed that it is the free chloride ions which cause damage to the passive layer on the rebars resulting in the corrosion of the steel.

A high percentage of the chloride present at the time of mixing the concrete may combine with the hydration products and will then be 'locked up' and not available for depassivation. On the other hand chloride entering the hardened concrete may be largely in the form of free ions and available to cause depassivation.

This differentiation between the effect on the steel of the combined and free chloride ions is important in connection with the cement type to be used for various conditions of exposure and is referred to in Chapter 2, section 2.2.3.

Chlorides in the concrete do not attack the concrete unless the concentration is very high. The formation of rust by chloride attack can cause cracking and spalling of the concrete due to the considerable increase in volume of the steel when it is converted into rust; an increase of between three to four times the original thickness of steel.

3.2.7 Stray electric currents

At the present time little is known/reported on the corrosion of steel rebars in concrete by stray electric currents. From a review of the very limited literature on the subject it appears that alternating current does not have any effect on rebars, and only direct current is likely to put the rebars at risk. Possible source of direct current is the installation of a cathodic protection system using an impressed direct current. Such systems must be very carefully designed by experienced firms. Some information of the use of cathodic protection is given in Chapter 6 and 9.

Prestressing steel is thought to be more at risk than ordinary reinforcement due probably to the risk of hydrogen embrittlement. Prestressing wires, due to their small diameter, are particularly vulnerable to the effect of corrosion.

However, high-quality, chloride-free concrete and mortar should be adequate to ensure that the passivating layer on the steel is intact.

3.3 DETERIORATION OF THE CONCRETE

3.3.1 Introduction

Deterioration of concrete mainly arises from two causes:

1. physical damage caused by impact, abrasion, freeze–thaw (frost), thermal shock, and high-velocity water; the cracking and spalling of concrete caused by the rusting of rebars has been dealt with in the previous section;

2. chemical attack, mainly on the cement paste, by external aggressive chemical compounds;
3. reaction between the alkalies in the hydrating cement and certain types of silicious aggregates; this is known as alkali-silica reaction.

3.4 PHYSICAL DAMAGE

3.4.1 Abrasion

Abrasion of the surface of concrete can occur from a variety of causes. On industrial floors by wheels from heavily loaded trolleys and heavy objects being dragged across the floor. On marine structures caused by sand and shingle being thrown against the concrete by heavy seas and gale force winds. To resist abrasion, the use of a high-quality concrete is required made with selected aggregates such as granite and flint gravel. Curing and method of finishing the concrete is also important. Reference should be made to Chapter 8, section 8.12.1.

Figure 3.1 Frost damage to large-diameter concrete pipe laid above ground in the Scottish Highlands.

3.4.2 Freeze–thaw

This type of damage is caused by the penetration of water into the surface layers of the concrete and then this moisture being converted into ice by sub-zero temperatures. Water expands on freezing, increasing in volume by about 10%, and this can cause disintegration of the surface layers of the concrete. See Figure 3.1.

Hydraulically pressed concrete slabs and paving blocks which possess high strength and low absorption will generally resist freeze thaw conditions, but for what may be termed 'normal' quality reinforced concrete, air entrainment should be used. See Chapter 2, section 2.6.5.

At the time of writing this book, there was no British Standard test for freeze–thaw, but a Draft European Standard was in preparation. The ASTM test is ASTM C.666: 1992: Standard Test Method for Rapid Freezing and Thawing.

3.4.3 Thermal shock

Damage to concrete by thermal shock is rare and occurs in only a few special cases such as the spillage onto concrete of liquified gases. The gases involved are those used in industry and include oxygen, nitrogen, methane, hydrogen and carbon dioxide. These gases liquify at the following approximate temperatures:

Oxygen	–183°C
Nitrogen	–196°C
Methane	–160°C
Hydrogen	–253°C
Carbon dioxide	–78°C

Generally, the damage is confined to floors and hard-standings and can be very severe. A mild form of thermal shock can occur if cold water is sprayed onto in situ concrete walls, cast in timber formwork during hot weather.

3.4.4 High-velocity water

The causes of damage to concrete by high-velocity water can be divided into three main categories:

1. cavitation;
2. water containing abrasive matter in suspension;
3. impact from a high-velocity jet.

3.4.5 Cavitation

The exact causes of cavitation have not been finally elucidated despite the considerable research and investigations which have taken place in various countries.

It is however clear that the degree of smoothness of the surface and the quality of the surface in terms of abrasion resistance play an important role as well as the actual velocity of the water.

No surface is absolutely smooth (even steel) and if the velocity of flow exceeds about 15 m/sec. there is a definite possibility that cavitation eddies will form at discontinuities on the surface and this will result in cavitation damage.

A simple and consequently incomplete explanation of this phenomenon is as follows.

If the absolute pressure at the points of surface irregularities approaches the vapour pressure of the water, minute bubbles will form and very quickly collapse. This collapse of the bubble walls can produce minute water jets having extremely high velocities and in turn this results in a series of hammerlike blows in extremely rapid succession. The effect is very destructive to concrete of the highest quality, and even to metals. See Figure 3.2.

Figure 3.2 Damage to concrete by cavitation in test rig. Courtesy, M.J. Kenn, Imperial College, London.

Further information on this subject is contained in the relevant papers listed in the Bibliography at the end of this chapter. Generally, cavitation damage has been found to occur to spillways, penstocks, aprons, energy dissipating basins and to syphons and tunnels in hydro-electric schemes.

There is no British Standard test for cavitation erosion, nor a European Standard. The ASTM test is G32-72: Standard Method of vibratory cavitation erosion test.

3.4.6 Water containing abrasive matter in suspension

High-quality concrete is very resistant to damage by abrasion from fast-flowing water containing grit or other abrasive material. The degree of damage caused will depend on a number of factors of which the following are the more important:

1. the quality of the concrete in terms of compressive strength cement content, the wear resistance of the fine and coarse aggregate;
2. the smoothness of the surface of the concrete;
3. the velocity of the flowing water;
4. the quantity of the grit carried, its particle shape, and its abrassive characteristics; the abrassive characteristics are not always obvious; for example china clay waste can be extremely abrassive;
5. the flow characteristics, i.e. whether continuous or intermittent, and the variation in the quantity of grit being transported.

3.4.7 Impact from a high-velocity water jet

The effect of a jet of water of very high velocity striking a concrete surface is to erode the cement paste, resulting in the loosening of the coarse and fine aggregate. The amount of damage done depends on the velocity of the water at the point of impact and the quality of the concrete in terms of compressive strength and cement content.

This effect is used in the cutting of concrete by high-velocity water discharged through specially designed nozzles; the pressure at the nozzle can vary from about 20 N/mm^2 to 40 N/mm^2.

3.5 CHEMICAL ATTACK ON CONCRETE

3.5.1 Introduction

Concrete is vulnerable to chemical attack by a wide range of chemical compounds in solution, and this applies to both Portland cement concrete and high alumina cement concrete. From a practical point of view, the chemicals which are aggressive to concrete can be divided into five main categories:

1. Acids – all;
2. ammonium compounds (some, not all);
3. magnesium compounds (some, not all);
4. sulphates – all;
5. other, including alkali hydroxides.

The intensity of attack depends on a number of factors, the principal ones being the chemical composition of the aggressive agent, the concentration, the pH and the porosity and permeability of the concrete, the type of cement used, and the contact time. Detailed consideration of the subject is outside the scope of this book but some general principles are set out below, and deal mainly with Portland cement concrete.

There are a vast number of chemical compounds used in industry and many of these are closely guarded trade secrets. However, some basic information is needed when considering the possible effect of a trade effluent on concrete sewer pipes. I understand that the Water Authorities are entitled to obtain information on the constituents of a trade effluent which is discharged or intended to be discharged to a public sewer.

3.5.2 Attack by acids

Both Portland cement and high alumina cement are highly alkaline (pH 13.0 and 11.8 respectively) and are therefore attacked by acidic solutions. Generally speaking, inorganic acids, such as the three set out below, are more aggressive than organic acids in the same concentrations:

- sulphuric acid (H_2SO_4);
- hydrochloric acid (HCl);
- nitric acid (HNO_3).

These are very aggressive and visible chemical action starts immediately the acid comes into contact with the cement paste, and with calcareous aggregates if used in the concrete/mortar (see Figure 3.3).

Of the organic acids, lactic acid is particularly aggressive to Portland cement concrete, but HAC is much less severely attacked. Acetic, citric and tartaric acids attack both Portland cement and HAC, but the rate of attack is slow compared with the inorganic acids listed above. According to F.M. Lea in *The Chemistry of Cement and Concrete*, 3rd edition, p. 663, organic acids of high molecular weight, such as oleic, stearic and palmitic acids have definite action on concrete.

They are present in the constituents of oils and fats and therefore industries where these compounds are used or prepared, should take appropriate action to protect concrete floors etc. where spillage can occur.

Figure 3.3 The effect of sulphuric acid on Portland cement concrete.

3.5.3 Ammonium compounds

Most ammonium compounds are aggressive to concrete; an exception is ammonium carbonate. The ammonium compounds used in the chemical fertilizer industry (sulphate, nitrate and super-phosphate) cause serious deterioration of concrete in a relatively short time; the actual time depends on the concentration and period of contact and whether the concrete is subject to abrasion (as in an industrial floor slab), and the quality of the concrete in terms of porosity, permeability, and cement content.

The aggressive ammonium compounds react with the hydration products in the Portland cement. According to F.M. Lea, in the *Magazine of Concrete Research,* Sept. 1965, the damage to concrete caused by ammonium sulphate is mainly due to the expansion arising from the formation of calcium sulpho-aluminate; while ammonium nitrate and ammonium chloride solutions act rather like dilute acids by reactions with lime in the hydrating cement past.

I have found that the addition of condensed silica fume to Portland cement concrete will increase its resistance to attack by ammonium-based fertilizers. See also Chapter 2, section 2.7.3.

High alumina cement is more resistent to ammonium compounds than Portland cement, but does suffer attack.

Ammonia vapours will attack moist concrete slowly.

3.5.4 Magnesium compounds

Magnesium chloride is aggressive to Portland cement concrete, but HAC is reported by F.M. Lea as being immune to attack. Magnesium sulphate is particularly aggressive to Portland cement but rather less so than ammonium sulphate. This attack is due to the reaction with hydrated calcium silicates in addition to its reaction with calcium hydroxide and calcium aluminates; this is known as double decomposition.

The extent of the damage and the speed at which it occurs depend on the same factors as those listed in the previous section on ammonium sulphate.

3.5.5 Sulphates

The degree of aggression depends in the first instance on the base of the sulphate compound and whether the solution is acidic.

The effect of ammonium and magnesium sulphates on concrete has already been dealt with in the previous sections.

Other common sulphates likely to come into contact with concrete are calcium sulphate (gypsum), sodium sulphate and potassium sulphate.

Calcium sulphate is fairly insoluble, as a saturated solution is formed at about 1100 ppm, while sodium and potassium sulphates are much more soluble. This is important as these salts have to be in solution in order to react with the hydration products of the cement.

The reaction of these sulphates when found in natural soils and ground waters is slow and the concentration relatively low. With well-compacted good-quality concrete, little damage is likely to occur even after many years. This is shown by the results of long-term tests (25 years) on concrete specimens buried in the subsoil at Northwick Park Hospital, near London and reported by the Building Research Establishment. See the Bibliography at the end of this chapter.

However, much higher concentrations can be present in the subsoil and ground water in industrial tips and in trade effluents and this would require special precautions to be taken to protect the concrete.

Concrete ground floor slabs and oversite concrete are vulnerable to sulphates in hardcore on which the concrete is laid, and a number of serious failures have been reported.

The sulphates listed above react with the C3A in the hydrating Portland cement and this reaction is expansive and the resulting compounds about double the solid volume of the reaction compounds.

It is this expansion which is largely responsible for the damage to the concrete.

BRE Digest[1] (1991) sets out in some detail recommendations for selection of cement types, minimum cement contents and maximum water/cement ratios needed to resist sulphate attack resulting from different concentrations of sulphates, including magnesium sulphate.

Sodium hydrogen sulphate ($NaHSO_4$) which is used in the treatment of swimming pool water is acidic and is aggressive to concrete.

Ferrous sulphate ($FeSO_4$) is used in water treatment and is also acidic and will attack concrete.

3.5.6 Chlorides

Solutions of sodium and potassium chlorides do not cause deterioration in good-quality concrete. However, a concentrated solution of calcium chloride can cause gradual deterioration to concrete, and this may occur in brine freezing plants due to spillage. As previously mentioned, chlorides will attack steel reinforcement.

Ammonium, aluminium, and magnesium chlorides in solution will attack concrete.

A further matter for consideration is that the build-up of salts in concrete due to alternate wetting and drying, can result in spalling of the concrete surface, for example in marine structures, see Chapter 9, Part 3.

3.5.7 Sodium hydroxide (caustic soda)

Solutions of caustic soda up to about 10%, will not adversely affect concrete, but higher concentrations, particularly at elevated temperatures, are likely to cause slow deterioration of concrete, with high alumina cement concrete being more vulnerable than Portland cement.

3.5.8 Distilled and demineralized water

It is surprising that very pure water should be aggressive to concrete, but this is due to the liquid's very high dissolving power. The main characteristics of this type of water are:

- very low calcium hardness (virtually zero);
- very low total dissolved solids (virtually zero);
- low alkalinity.

See Figure 3.4.

If the water contains dissolved carbon dioxide, thus bringing the pH down to below the neutral point of 7.0, this will increase the potential for attack. The water also has a negative Langelier Index which means that it is 'lime dissolving'. The Langelier Index is used in the control of treatment for boiler feed water, but it is also useful in the field of the treatment of soft moorland water stored or conveyed in concrete tanks or pipelines. See section 3.5.9 below.

Figure 3.4 Damage to concrete by hot distilled water. Courtesy, Dr D. Higgins.

An ISO document, No. N18E dated February 1983, gives the classification for Langelier Index for water shown in Table 3.1.

Table 3.1

Langelier Index	*Water classification*
–2.0 and lower	Highly aggressive
0.0 to –2.0	Moderately aggressive
Any positive value	Not aggressive

3.5.9 Moorland waters

Waters from upland gathering grounds can have a long-term aggressive effect on concrete used to store and convey such water. The characteristics of such waters are:

- low total dissolved solids (TDS);
- low total hardness;

- low pH (usually below 4.5);
- low alkalinity;
- organic and other acids in dilute solution; sometimes sulphuric/sulphurous acids are present, derived from the breakdown of organic matter in peat and marshy ground.
- usually a negative Langelier Index;

See Figure 3.5.

Figure 3.5 The effect of soft moorland water of low pH on concrete water channels.

3.5.10 Sea water

Generally this is not aggressive to dense good-quality concrete. The concentration of dissolved salts, mainly chlorides and sulphates vary. In Atlantic water there are about 18 000 ppm of chlorides and 2500 ppm of sulphates. In the Red Sea and the Persian/Arabian Gulf the concentrations are about 25% higher. It is generally not recommended to use sulphate resisting Portland cement for marine structures due to its low C3A content; see comments earlier in this chapter in section 3.2.6. Sea water in tidal estuaries may become seriously contaminated by trade effluents and this should be taken into account. Further information on the effect of sea water on concrete is given in Chapter 9, Part 3.

3.5.11 Sewage–domestic and trade effluents

Ordinary domestic sewage is not aggressive to Portland cement concrete. However, when sewage becomes septic due to relatively high temperatures and/or retention in sumps of pumping stations or long lengths of sewer with flat gradients and low velocity, the hydrogen sulphide–sulphuric acid cycle can occur with disastrous results to any unprotected concrete above the top water line. This type of attack is rare in the UK, but has occurred with serious consequences in Australia, S. Africa, the countries bordering the Arabian Gulf and parts of the USA.

The process is known as hydrogen sulphide corrosion. The main feature of this attack is that it occurs above the waterline in sewers, the sumps of pumping stations, manholes and sludge digestion tanks.

The basic characteristics of this attack are as follows.

1. Sulphides must be present in solution in the sewage. The sulphides may originate from trade wastes or may be formed by anaerobic bacteria within the sewage.
2. A significant amount of hydrogen sulphide gas (H_2S) must escape from the sewage and be present in the air space above the sewage.
3. The H_2S is converted to sulphuric acid, mainly by aerobic bacteria in the air space which has a high moisture content.
4. The sulphuric acid thus formed attacks the cement paste in the concrete. The decomposition of the cement paste results in the destruction of the concrete.

Some comments on repair of concrete damaged in this way are given in Chapter 9, section 9.4.6.

3.5.12 Various compounds

3.5.12.1 Fruit and vegetable juices

These contain organic acids and sugar and are therefore aggressive to Portland cement concrete. HAC concrete is rather more resistant. These substances usually originate in food manufacturing plants and occur as spillage. The degree of attack depends mainly on the contact time between the spillage and the concrete, i.e. how quickly the spillage is cleaned up.

3.5.12.2 Milk and dairy products

Fresh milk and similar dairy products do not attack concrete, but when they become stale (ferment), lactic acid is formed which is aggressive to

Portland cement. The attack can be serious unless the fermentation product is quickly washed away. Rapid and thorough washing is essential to reduce/mitigate the attack.

3.5.12.3 Sugar

Sugar will attack concrete slowly, but if it penetrates into the concrete through cracks and/or an abraided surface, the rate of attack increases.

3.5.12.4 Petroleum oils

These products are generally not aggressive to Portland cement unless the acid and sulphur contents are high. Lubricating oils used in the car industry contain corrosion inhibitors and so spillage will not cause deterioration of a concrete floor. The staining is virtually impossible to remove as these oils have high penetrating properties.

3.5.12.5 Urea

There are differences of opinion as to whether the deterioration of concrete in contact with urea is due to chemical attack or whether it is a physical phenomenon arising from the growth of crystals in the pores of the concrete which disintegrate the surface layers (as with frost attack). From literature examined, it appears that the consensus of informed opinion is that there is no chemical reaction between the urea and the hydration products of the Portland Cement.

Urea is used in the chemical fertilizer industry, and to some extent as a de-icing salt on roads and airfield runways.

The principal de-icing salt is sodium chloride which is very aggressive to ferrous metals (steel reinforcement and the bodies of motor vehicles). However, work in the US by Forbes, Stewart and Spellman showed that urea is almost as aggressive to ferrous metals as sodium chloride.

I suggest that urea can cause deterioration of concrete and have seen surface damage to concrete floors in buildings used for the storage of urea in the **chemical** fertilizer industry.

3.5.13 Alkali–silica reaction

The deterioration is caused by the reaction between alkalis in the concrete and certain type of siliceous aggregates.

ASR was first reported in 1940 in the United States. Since then it has been identified as the cause of cracking and expansion of exposed concrete in many countries.

The first confirmed case in the UK was in 1976 in the concrete of a dam in the Channel Islands. There are differences of opinion as to the present (1996) number of confirmed cases of ASR. The differences of opinion on the actual deterioration of concrete affected by ASR, in terms of load-carrying capacity, frost resistance and increased risk of rebar corrosion, are even greater.

Agreement is general that the effects of ASR are long term and about five years is the minimum period required for any visible signs to appear (see Figure 3.6.).

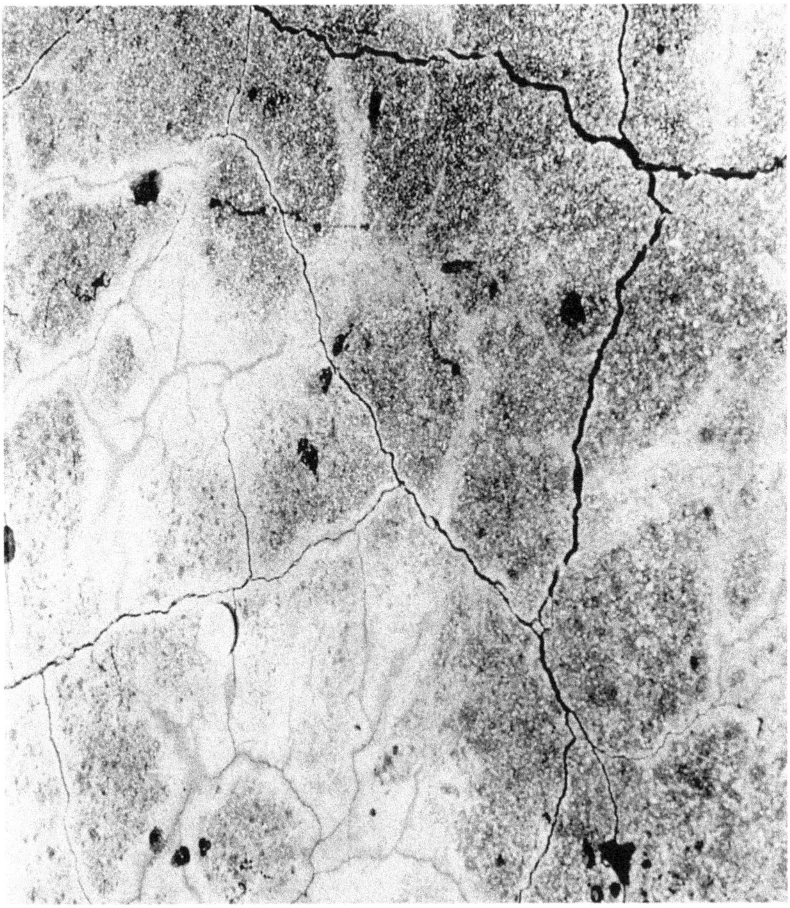

Figure 3.6 Damage to concrete showing 'map' cracking typical of alkali–silica reaction. Crown Copyright; reproduced from *British Research Establishment Digest*, **258**, by permission of the Controller of HMSO.

A great deal has been written about ASR and a selected number of papers on the subject are included in the Bibliography at the end of this chapter.

3.5.13.1 Summary of ASR problem

Some effects of ASR are visible, but these are by no means conclusive:

1. Mat-type cracking caused by differential expansion of the concrete within a pour or member. This crack pattern is modified by the restraint set up by the presence of reinforcement, and is sometimes mistaken for sulphate attack.
2. The formation of a gel around aggregate particles and in cracks.

A comprehensive examination by an experienced testing laboratory is essential.

As stated above, ASR arises from chemical reaction between alkalis in the concrete and certain types of siliceous aggregates. The alkalis generally originate in the cement and are present in the pore fluid. This reaction results in the formation of an alkali–silica gel, and this gel in contact with water expands and causes visible cracking.

The situation is complicated by the fact that, with aggregates in the UK, there is a maximum percentage of reactive silica, beyond which expansion decreases. This maximum amount (the 'pessimum') varies from one type of aggregate to another. The main precautions which can be taken to avoid ASR damage is to use a low alkali cement, and take steps to prevent solutions of alkalis coming into contact and penetrating the concrete.

The alkali content of cement is expressed as 'equivalent sodium oxide' (Na_20).

At the time of writing this book the recommended limit for equivalent sodium oxide in concrete is 3 kg/m^3 of concrete.

A report by the Mineral Industry Research Organization (MIRO) dated May 1992, recommended that, based on their research, the 3 kg limit could safely be raised to 4 kg/m^3.

On the assumption that alkalis do not find their way into the concrete after casting, ASR can be avoided by ensuring that the alkali content of the concrete does not exceed 3 kg/m^3 (or 4 kg/m^3 if the MIRO recommendations are accepted); aggregates with a record of satisfactory performance can be used. It is advantageous if saturation of the concrete can be prevented.

3.6 REFERENCE

1. Building Research Establishment (1991) Sulphate and acid resistance of concrete in the ground, *Digest*, **363**, 8.

3.7 FURTHER READING

Introduction

British Standards Institution (1986) *Guide to Building Maintenance Management*, BS 8210.

British Standards Institution (1993) *Guide to the Durability of Buildings, Building Elements, Products and Components* BS 7543.

Building Research Establishment (1991) Structural Appraisal of Existing Buildings for Change of Use; *Digest*, **366**, Oct.

Carbonation

British Cement Association (1994) Avoidance of carbonation induced corrosion; *Bulletin* **20**, Sept., 4, 5.

Currie, R.T. (1986) *Carbonation Depths in Structural Quality Concrete*, Building Research Estab. Report, p. 19.

Chlorides

Bamforth, P. (1994) Admitting that chlorides are admitted, *Concrete*, Dec. 18–20.

British Standards Institution (1985) The Structural Use of Concrete BS 8110: Parts 1 and 2.

British Standards Institution (1991) Concrete, BS 5328 Parts 1–4.

Spooner, D.C. (1995) Chloride attack, *Concrete*, Mar./April, 5.

Sulphates

Building Research Estab. (1991) Sulphate and acid resistance of concrete in the ground, *Digest*, **363**, 8.

Building Research Estab. (1992) *Sulphate Resistance of Buried Concrete* (Northwick Park: *Results after 15 years*, p. 150.

Building Research Estab. (1983) Hardcore, *Digest*, **276**, Aug. 4.

Chemical attack on concrete

American Concrete Institute (1979, rev. 1985) *A Guide to the Use of Waterproofing, Damp-proofing, Protective and Decorative Barrier Systems for Concrete*, ACI., 515., 1R–79, 43.

Building Research Establishment (1993) *Effects of Alkali–silica Reaction on Concrete Foundations*, BRE Information Paper IP16/93, Oct., 4.

Building Research Establishment (1991) Alkali aggregate reactions in concrete, *Digest*, **330**, 8.

CENTC104/SC1/TG9 (1995) *Regional Specifications and Recommendations for the Avoidance of Damaging Alkali–Silica Reactions in Concrete*, p. 5.

Hobbs, D. (1992) Deleterious alkali–silica reactivity of a number of UK aggregates, *Concrete*, May/June, 64–70.

Lea, F.M. (1970) *The Chemistry of Cement and Concrete*, 3rd edn, Edward Arnold (Publishers) Ltd, p. 727.

Sims, I. (1992) The assessment of concrete for ASR, *Concrete*, Mar./April, 42–6.

4

Investigation and diagnosis of defects in reinforced concrete

4.1 INTRODUCTION

Chapter 3 has detailed the principal factors which cause deterioration of both the reinforcement and the concrete in reinforced concrete structures.

This chapter is intended to cover the main factors involved in the investigation and diagnosis of defects and deterioration of concrete structures generally. Chapters 6, 8 and 9 deal with specific problems relating to floors and roofs, fire damaged concrete, highway bridges, silos and bunkers, liquid retaining structures, water excluding structures and marine structures. This results in a certain amount of duplication.

4.2 GENERAL OUTLINE OF THE PROCEDURE

Following the initial request from the client, the procedure adopted and the form of the report can be as set out below, but it will of course vary, depending on the type of structure involved and the magnitude of the deterioration.

4.2.1 Initial discussions and preliminary inspection

1. Arrangements should be made for an early discussion with the client and a visit to site, to enable an assessment of the problem to be made. The client should at this stage be asked by the engineer to issue him with a clear brief for the investigation he requires to be undertaken, unless of course he has already done so.
2. The engineer should try to obtain as much information as possible about the structure such as:
   (a) year of construction;
   (b) details of construction including drawings;

(c) present use, any previous changes of use, any proposals for future change in use;
(d) if the structure was constructed after the Second World War, the firms involved in the design and construction;
(e) when deterioration was first noticed and whether any repairs had been carried out, and if so, the details and approximate date(s) of execution.

3. The preliminary site visit and any information available (as set out in (2) above) will often indicate the possible underlying causes of the deterioration and thus enable the engineer to draw up proposals for a more detailed investigation, and arrangements for sampling and testing. Some recommendations for the preliminary investigation are given in section 4.3.

4.2.2 Detailed inspection, sampling and testing

1. The sampling and testing would necessitate the engagement of a commercial testing laboratory and the provision of access equipment. Detailed recommendations are given later in sections 4.4 and 4.5.
2. The interpretation of the results of the sampling and testing should be the responsibility of the engineer and would form an essential part of the report to his client.

4.2.3 The engineer's report to the client

On the completion of the investigations the engineer would prepare a detailed report to the client, setting out the instructions he received, a description of the investigations, and recommendations for remedial work.

Detailed suggestions for the form of the report are given in section 4.8.

4.3 THE PRELIMINARY INSPECTION

Ideally, the information which is listed in 4.2.1 (2) should be in the hands of the engineer before he makes his initial inspection but this is very seldom the case. Such information is hard to obtain and in many cases is just not available. The engineer then has to make his inspection on the basis of the client's complaint, such as 'pieces of concrete falling off', 'cracking', 'rust staining', 'water penetration' etc.

Depending on the type of structure and the alleged defects, the engineer should take with him: binoculars, a rebound hammer, bolster and heavy hammer, simple means for approximate measurement of crack widths, an electro-magnetic cover meter, a good camera, and a spray bottle containing phenolphthalein for the detection of depth of carbonation.

The absence of access for external inspection above ground level can impose severe restrictions on the amount of investigation possible at this stage, and therefore it may not be practicable to use all the items listed.

The object of the inspection is to obtain a general picture so that practical decisions can be taken on the following matters:

1. the need or otherwise for specific information; for example, if structural (load-bearing) deficiency is suspected, or the client has indicated a wish for change of use which would increase the floor loading, an inspection of the original drawings and calculations can save a considerable amount of time, disruption, and expense.
2. the preparation of a sampling and testing programme, including selection of means of access (scaffolding, cradles etc.).

If at all possible, a number of samples of the concrete should be taken and the locations noted, and depth of carbonation at these sampling locations should be recorded. The approximate location, length and width of a number of selected cracks should be noted, and as far as possible an assessment of the probable cause.

The samples should be visually examined for assessment of degree of compaction, and type and size of aggregate, and analysed to determine cement content, and chloride content (expressed as percentage by mass of the cement).

Samples of concrete can usually be taken by the careful use of a heavy hammer at locations where cracking and spalling has occurred, and these can be used for an initial determination of cement content, and of chlorides and sulphates as a percentage of the cement.

For the initial determination of the cement content, if pieces of concrete are not available, samples can be taken by dry drilling, using a percussion type masonry drill; the powder from the first 5 mm depth should be discarded.

4.4 DETAILED INSPECTION, SAMPLING AND TESTING

4.4.1 Introduction

For the purpose of the recommendations which follow, it is assumed that the defects are non-structural, that is, that in its present state the building is structurally sound.

The engineer, in consultation with an experienced testing laboratory should prepare a clear brief for the laboratory, detailing the location and type of samples and the amount of testing to be carried out. This brief may have to be revised in the course of the investigation.

The selection of the laboratory is important and it should be a laboratory accredited by the United Kingdom Accreditation Service (UKAS). This should ensure that the testing is carried out strictly in accordance with standard test methods. Should a standard test method not be available to provide desired information, then the laboratory in consultation with the engineer would have to devise a practical test.

The principal British Standard for testing concrete is BS 1881 and is referred to as appropriate in this chapter.

The location, number and type of samples should enable the testing to be representative of the members under investigation, and the testing must be relevant to the information required. The sampling, testing, and provision of access equipment is expensive, but is an important part of the overall investigation.

The preparation of the sampling and testing programme requires considerable experience and a large measure of common sense.

4.4.2 The number, location and type of samples

The number of samples will depend mainly on the number of components (beams, columns, wall panels etc.) under investigation. It is usual to take a sample from at least 10% of the units, with a minimum of four if there are less than 20 units. If the results of the tests show wide variations, then the range of sampling would have to be correspondingly increased.

In deciding on the amount of testing, it is necessary to consider carefully what information is required from the test results and how this information can be used to determine the cause of the defects.

4.4.3 Depth of carbonation

This test is important due to the relationship between depth of carbonation and possible corrosion of the rebars. Tests for depth of carbonation are best carried out on site using an indicator on freshly exposed surfaces. These indicators are acid-based and give different colours according to the alkalinity or pH of the exposed surface. The most commonly used indicator is phenolphthalein which turns a purple-red colour on uncarbonated surfaces (pH above about 10); the carbonated surfaces retain their original colour. The depth of carbonation can then be seen and measured.

A record should be prepared showing location of these tests and depth of carbonation, supplemented by a few colour photographs.

Good-quality dense concrete carbonates very slowly, so that even after a period of some 50 years, carbonation is unlikely to penetrate to a greater depth than 7–10 mm. On the other hand, permeable concrete may carbonate to a depth of 25 mm in ten years or less. The rate of carbonation generally decreases with time as the formation of calcium carbonate acts as a pore filler.

Carbonation does not adversely affect the strength of the concrete and tends to reduce permeability to the passage of moisture. Further information on carbonation is given in Chapter 3, section 3.2.5.

4.4.4 Type and grading of aggregate

It is desirable to determine the type of aggregate used. With calcareous aggregates, the accuracy of cement determination is reduced.

The assessment of the grading of aggregates from samples of hardened concrete can only be approximate due to the test procedure (see BS 1881: Part 124) and should only be used to help supplement the results of other tests.

4.4.5 Cement content of the concrete

The cement content should be determined in accordance with BS 1881: Part 124, which gives practical advice on sampling and detailed recommendations for analysis.

Pieces of concrete knocked off the members can also be used for cement determination, provided they are whole, do not contain cracks and weigh at least 1 kg. The aim should be to obtain a combined sample weighing several kg.

For a more accurate determination of the cement content of the concrete, 50 mm dia. cores can be drilled at selected locations, care being taken with the aid of a cover meter to avoid rebars.

The taking of cores should be kept to a minimum but they do provide a good picture of the concrete and are needed for a proper assessment of voidage and standard of compaction.

Care and experience is needed in the interpretation of the results. With chemical analysis to establish the cement content and thus the mix proportions of the concrete, there is considerable reduction in the sample size. Only about 5 gms of powdered concrete/mortar is analysed from a combined sample of several kg. Thus the preparation of the final sample is of crucial importance to ensure that the sample analysed is truly representative of the combined sample and of the original concrete.

There is testing variability on samples tested within one laboratory and between laboratories, even though they may be NAMAS accredited.

Concrete Society Technical Report 32: Analysis of Hardened Concrete, suggests sampling variability as + or -25 kg/m^3 and testing variability as + or -30 kg/m^3 making a combined variation of + or -40 kg/m^3 (the square root of $25^2 + 30^2$).

The cement content of good structural quality concrete is likely to be about 300 kg/m^3, or 12–14% by mass of the concrete.

4.4.6 Cement type

In practice, tests to determine cement type are only likely to be necessary in cases where it is suspected that high alumina cement may have been used, or where sulphate resisting Portland cement was specified and there is some reason to believe OPC was used. HAC is appreciably darker in colour than Portland cements, but the colour of some sulphate resisting Portland cements may approach that of HAC.

Some information of the principal characteristics of HAC concrete have been given in Chapter 2, section 2.3.

Due to the major differences between Portland cement and HAC, if there is any doubt about the possible use of HAC concrete then the necessary tests should be carried out. In the 1970s, there was great interest in the use of HAC concrete (due to a number of failures) and BRE developed a rapid chemical test using relatively simple apparatus and procedures; details are given in a BRE information Sheet, IS.15/74 (reprinted Feb. 1975). If this test gives positive results, then an adequate number of samples of the concrete should be taken and passed to an experienced testing laboratory. HAC concrete was mostly used for precast units, especially prestressed beams and floor units.

4.4.7 Chloride content of the concrete

The chloride content can be determined by the method set out in BS 1881: Part 124: Section 10.2. In all cases where corrosion of rebars has occurred, the concrete must be checked for chloride content. The permitted limits of chloride content of concrete are set out in BS 5328: 1991, Part 2, Table 8. The chloride content is expressed as a percentage of chloride ion by mass of cement. For prestressed concrete this is limited to 0.10%; for concrete made with SRPC, the limit is 0.20% and for other Portland cements the limit is 0.40%. The reason for the reduced percentage in SRPC concrete is discussed in Chapter 3, section 3.2.6.

4.4.8 Sulphate content of the concrete

The sulphate content of concrete can be determined by the method set out in section 10.3 of BS 1881: Part 124.

Sulphate is present in Portland cements, as gypsum is added during the manufacture to control the set, but is limited to about 3% expressed as SO_3 by mass of the cement. There are no tests available which can determine a specific limit to the 'safe' sulphate content. However, a limit of 4% by mass of the cement is considered reasonable.

I consider that testing for sulphate is generally not necessary unless there is evidence which suggests that sulphate attack has taken place, as may occur in foundations and ground-supported floor slabs, and ground beams.

4.4.9 Assessment of voids and compaction of the concrete

In assessing the quality of the concrete, a reasonable determination of the standard of compaction is desirable. Visual examination of pieces of concrete can give a general idea, but where information on compaction is considered essential, 50 mm dia. cores should be cut and examined as described in BS 1881, Part 120. Reference can also be made to the 1987 edition of Concrete Society Technical Report 11. Ultra-Sonic Pulse Velocity and Impulse Radar which are briefly described in 4.4.10.5 and 4.4.10.6 can also be used for detecting voids and honeycombing.

4.4.9.1 Honeycombed concrete

A suitable definition of honeycombed concrete would be:

> Concrete which has large interconnected voids resulting from loss, or original lack of, cement, aggregate fines and water.

Generally, it is due to the loss of cement paste through leaks in the formwork, but can arise from inadequate mixing, careless placing and compacting, and congested reinforcement. Cast in situ piles are particularly prone to honeycombing.

My experience is that when this defect exists it can usually be seen when formwork is removed, but in massive units it can occur below the surface of the member. In the latter location it is only likely to be detected by a UPV or Radar survey. Some comments on methods of repair are given in Chapter 5, section 5.3.9 and Chapter 9, section 9.4.5.

4.4.10 Additional tests on the concrete

It may be desirable to carry out some of the following additional tests:

1. original water content (from which the w/c ratio can be calculated) (BS 1881: Part 124: Section 7);
2. the water absorption (BS 1881: Part 122);

3. initial surface absorption (BS 1881: Part 201: Sect. 2.8);
4. rebound hammer tests for surface hardness (BS 1881: Part 202);
5. ultra-sonic pulse velocity tests (BS 1881: Part 203);
6. impulse radar.

Comments on each of the above tests are given in the following sub-sections, 4.4.10.1 to 4.4.10.6.

It should be noted that some 24 non-destructive tests are listed in Section Two of BS 1881: Part 201: Guide to the Use of Non-destructive Methods of Test for Hardened Concrete. Some of these are very useful for special investigations, and where appropriate, are mentioned in other chapters in this book.

I would again emphasize that testing should only be carried out when the information given by the tests is needed for diagnosis of the cause of the defects.

4.4.10.1 Original water content

I have not found a need for this test, but I am aware that some engineers use it.

The 'original water content' means the amount of water in the concrete at the time of setting. BS 1881: Part 124 makes it clear that there are definite limitations to its acceptability, namely:

1. the concrete being tested should not be older than five years;
2. the sample being tested must not be damaged in any way, either physically or chemically;
3. the concrete must be well compacted, and not air-entrained;
4. the value of the test is compromised if the concrete contains hydraulic binders other than Portland cement;
5. concrete made with porous aggregates introduces unacceptable errors.

4.4.10.2 Water absorption

BS 1881: Part 122 requires that the test specimens should be cores, which are dried in an oven for 72 hours at a temperature of 105°C, then allowed to cool for 24 hours, weighed, and then immersed in water for 30 mins or other prescribed period and weighed again.

The question mark over a decision to carry out this test is: what is an acceptable absorption for the concrete under investigation? There is little justification for carrying out a test unless there is an acceptable criterion against which the test results can be compared. The only authoritative figure I have found relates to concrete pipes for sewers, BS 5911: Part 1, clause 25.2, which requires the maximum absorption after 30 minutes should not exceed 3.6% and after 24 hours should not exceed 6.5%.

4.4.10.3 Initial surface absorption test (ISAT)

This measures the rate of water absorption by a concrete surface under a constant applied head. The rate will decrease with time and so measurements have to be taken at specified time intervals from the start of the test.

The test results are affected by many factors which are set out in BS 1881: Part 208. One of the most important is the moisture content of the surface layers of the concrete at the time of the test. I feel that this factor alone makes the test difficult to apply realistically to external in situ concrete.

The test is used mainly on precast concrete units as a quality control test. I have not seen reports of this test having been used for the type of investigations described here. The basic difficulty with using this test on existing precast units is to decide what the actual results can be compared with. This problem and others are set out and discussed in BS 1881: Part 201: Sect. 2.8 and BS 1881: Part 208.

It should be noted that BS 1217: Cast Stone sets ISAT limits for two grades of product, Grade A and Grade B. For Grade A, the ISA must not exceed 0.25 mL/(m^2s) at 10 min. and must not exceed 0.10 mL/(m^2s) at 1 hour.

4.4.10.4 Rebound hammer test

The rebound hammer (sometimes referred to as the Schmidt hammer) essentially measures the surface hardness of concrete, but the results are often used to indicate the variability of concrete in a unit or member. I have found the rebound hammer very useful in assessing whether further testing by taking cores is justified for the purpose of the investigation. It should not be used as a 'pass or fail' test.

For example, surface hardness is important in industrial floors and the rebound hammer is a very useful tool in such an investigation. A minimum of 15 readings per bay or unit should be taken and I suggest that the highest and lowest reading should be discarded. I have found that because this is a site test which can be easily and quickly carried out, there is a tendency to use it for the assessment of the compressive strength of the concrete. This can be considered as valid if the actual strength of the concrete in other similar parts of the structure are known. BS 1881: Part 202 lists five factors which influence the relationship between the surface hardness and the compressive strength. However, even under very favourable conditions the test should only be used as an indication of the probable compressive strength.

4.4.10.5 *Ultrasonic pulse velocity tests*

BS 1881: Part 201, Sect. 2.13, gives a useful summary of the test and details how it should be carried out. The principal uses of the test for the purpose of detecting defects in concrete are:

1. locating the presence and approximate extent of cracks and voids, particularly in grouting in prestressing ducts;
2. checking on uniformity/homogeneity of the concrete in or between members or different parts of the structure.

The effective use of upv for detecting defects in concrete requires considerable experience, particularly in the correct interpretation of results. It is a sophisticated method of test and expensive but for large and/or complex jobs it can be an essential part of the investigation.

4.4.10.6 *Radar (impulse radar)*

This is also known as pulsed radio echo sounding. It is briefly described in BS 1881: Part 201, Section 2.12, but at present there is no British Standard for this technique.

Impulse radar was first introduced in the early 1980s and has been used for locating underground services and for the non-destructive testing of concrete. In the latter field its main use is in the detection and location of voids, delamination at the interface between two materials (e.g. a concrete slab and a screed or topping), and reinforcing bars.

The equipment used requires very experienced operators and specialist interpretation of the test results. For further information see Chapter 6, section 6.2.4.

4.5 TESTS FOR THE DETECTION AND DIAGNOSIS OF REINFORCEMENT CORROSION

4.5.1 General considerations

The tests normally used are:

1. determination of depth of carbonation;
2. examination of cracks and crack pattern;
3. cover-meter survey;
4. the 'half-cell' test to determine the electrical potential of embedded rebars and thus the approximate degree of corrosion;
5. the determination of loss of section of rebars due to corrosion;
6. radiography to obtain photographs of the interior of concrete members which show the rebars.

4.5.2 Depth of carbonation

This test has been described under testing the concrete (section 4.4.3) to which the reader is referred. The reasons why carbonation of concrete is important in investigating the corrosion of reinforcement are discussed in Chapter 3, section 3.2.5.

4.5.3 Cracks and crack pattern

Section 4.6 discusses in some detail the factors involved in cracking in reinforced concrete, as the presence of cracks can be a major factor in the corrosion of the reinforcement. If cracks extend down to the rebars they provide access for moisture, carbon dioxide, sulphur dioxide and oxygen, and aggressive liquids, and this will result in corrosion of the steel reinforcement.

Reinforcement can and does corrode without being initiated by cracks, e.g. by porous concrete, inadequate depth of cover, presence of chlorides in the original concrete mix, and carbonation of the concrete in contact with the steel.

The rust which occupies about 3–5 times the volume of the original meter exerts considerable pressure on the concrete. This causes, first, the formation of fine cracks which gradually increase in width, and then pieces of concrete start to be debonded and eventually break off (spalling).

It is therefore necessary to carefully examine all exposed concrete for the presence of these fine cracks, generally in line with main and/or secondary reinforcement.

These areas (defined by the fine cracks) should be tested by a medium weight hammer; this is known as hammer testing. This hammer testing will show where corrosion is sufficiently far advanced to crack and partially debond the concrete cover over the rebars.

4.5.4 Cover-meter surveys

The use of a cover-meter to check the depth of concrete cover to the reinforcement is an essential part of any investigation into deterioration of reinforced concrete. However, a visual inspection can be very useful in detecting the approximate extent to which corrosion of rebars has caused cracking and spalling of the concrete cover, as this physical damage to the concrete is usually accompanied by rust staining of the surface.

It must be mentioned that rust stains can also be caused by the presence of pyrites in the aggregate, so that the existence of rust stains alone is insufficient to conclude that rebar corrosion has occurred. However, localized corrosion can cause pitting in the rebars which may not show on the surface of the concrete. This is one of the reasons why the half-cell is used – it is described in section 4.5.5.

Detailed recommendations for the use of electromagnetic cover-meters are contained in BS 1881: Part 204. The cover-meter usually consists of a search head, a battery, a meter showing depth of cover, and a cable.

A correctly calibrated cover-meter should indicate the cover (distance from the surface of the concrete to surface of the rebar), to an accuracy of + or –2 mm or + or –5% whichever is the greater, over the range given by the manufacturer. However, BS 1881: Part 204, Section 8, quite rightly emphasizes that the results on the average site, when used by an average operator, would require a larger tolerance, namely + or –5 mm or + or –15%, whichever is the greater for covers less than 100 mm.

It is important that the calibration of the cover-meter be checked at regular intervals, either as recommended by the manufacturer or by means of a specially made block of concrete containing a reinforcing bar which projects not less than 100 mm at both ends of the block.

Drawings should be prepared showing each member surveyed with the cover-meter readings plotted on. The number of readings and their locations will depend on the circumstances of each case; readings closer than 300 mm centres are unlikely to be needed.

The recommendations for nominal cover to reinforcement are contained in Codes of Practice and Standards and are related to exposure conditions and the quality of the concrete in terms of cement content and water/cement ratio.

The relevant exposure conditions are listed in Tables 5 and 6, BS 5328: Concrete: Part 1: Guide to Specifying Concrete, and range from 'mild' to 'most severe' and 'abrasive'. This is discussed later in this chapter under 'Diagnosis', Section 4.7.

4.5.5 The half-cell potential measurements

This method appears to have been developed in the USA in the 1960s and is covered by ASTM Specification C876-80: Standard Test method for Half-cell Potentials of Reinforcing Steel in Concrete. There is no British Standard for this technique, but it is described in BS 1881: Part 201, Section 2.4.

It measures the potential of an embedded rebar relative to a half-cell, and consists of a reservoir containing a saturated solution of copper sulphate ($CuSO_4$), and secured centrally within the container is a copper rod connected by an electric lead to a high-impedance voltmeter. In one end of the container is a sponge plug which remains continuously saturated with the copper sulphate solution. The saturated sponge can be considered as the search head of the apparatus. The voltmeter is connected to a rebar in the concrete.

The surface of the unit to be examined is divided into areas of about 300 mm × 300 mm and this grid is marked in chalk on the unit.

Immediately prior to the test the surface of the unit is sprayed with water. The search head is then placed in contact with the damp surface of the concrete in the centre of each grid and readings recorded on to a drawing. An alternative method of recording is to move the search head about so as to locate lines of equal potential (contour lines). Figure 4.1 shows a 'contour' map.

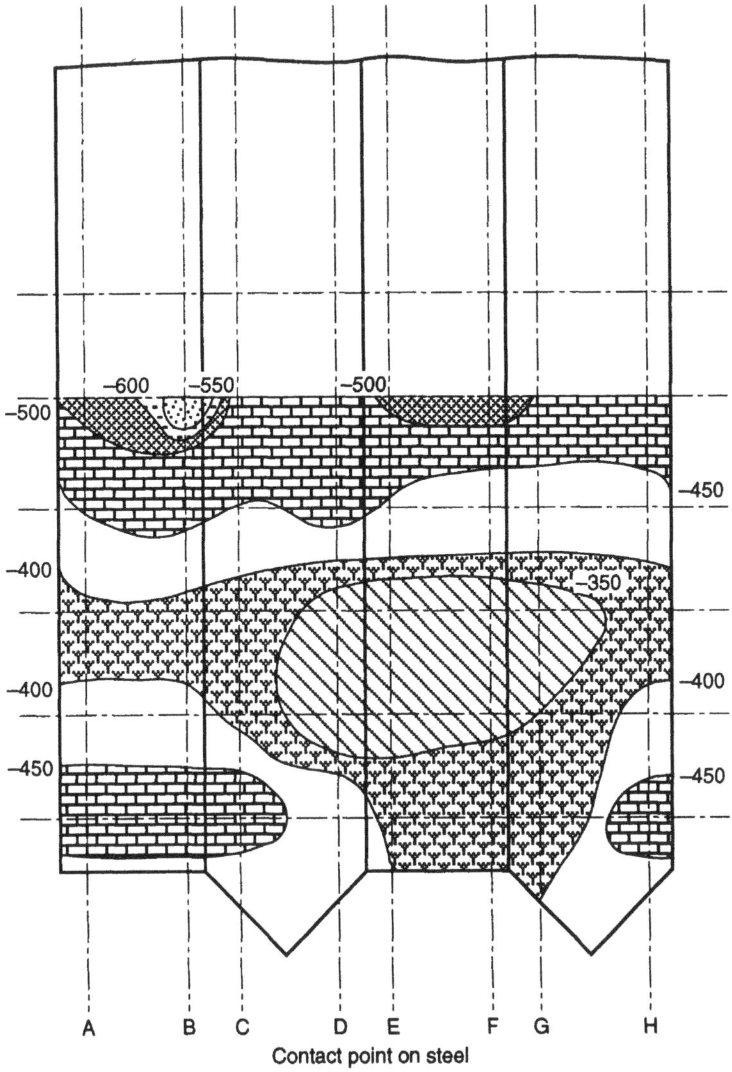

Figure 4.1 Half-cell interpretation map: contact point on main reinforcement below gunite. Courtesy, Technotrade Ltd.

It is important to note that the half-cell does not indicate the rate of corrosion, but the readings do indicate the intensity of the corrosion which is taking place.

The implications of the readings are as follows:

1. for potentials less negative than –200 mv there is a 90% probability that corrosion is not taking place;
2. for readings between –200 mv and –350 mv, there is a 50% probability of corrosion;
3. for readings numerically greater than –350 mv, there is a 90% probability that corrosion is taking place.

There are a large number of factors which influence the readings, the principal ones being the moisture content of the concrete, the presence of salts in the concrete and significant variations in the carbonation of the surface of the concrete.

4.5.6 Determination of loss of section of rebars due to corrosion

The obvious method is to expose selected sections of corroded rebars and make direct measurements of the loss of section due to corrosion. Due to the significant expansion of steel when it is converted into rust, a 'lot' of rust may result from comparatively small loss of original steel and consequently a significant reduction in the bar diameter. However, this test is important because pitting corrosion can be potentially dangerous if there is significant reduction in bar diameter at the pits.

For continuous monitoring for detection of corrosion of rebars after completion of repairs, see Chapter 6, section 6.8.

4.5.7 Radiography

This technique enables a photograph to be taken showing details of the inside of a concrete member.

It is used to determine the location and size of reinforcement, to check for the existence of voids and areas of poor compaction in concrete where other NDT methods are not suitable, such as massive old concrete structural units when being assessed for structural safety. It is also used for checking for voids in the grouting in prestressing ducts.

BS 1881: Part 205 gives detailed information/recommendations for the radiographic inspection of concrete.

The method uses sources of gamma rays for concrete up to about 500 mm thick and, above this thickness, the Standard recommends the use of high-energy X rays.

Due to the inherent danger of using gamma rays and high energy X rays, the use of this method is confined to investigations which justify the cost of the special precautions which have to be taken.

4.6 CRACKING IN REINFORCED CONCRETE STRUCTURES

4.6.1 General considerations

It is now necessary to discuss in more detail the formation of cracks in reinforced concrete. This subject has been mentioned briefly in Chapter 3, section 3.2.2.

A careful and informed examination of the cracks will provide valuable information which should generally indicate the likely cause. Such factors as crack position, pattern, direction (vertical horizontal or inclined) in relation to the main reinforcement and secondary reinforcement, the width of the cracks. Some indication of their age can be obtained by reference to broken (spalled) edges of the cracks and to the accumulation of dirt in the cracks. The width of the cracks will be found to vary considerably along their length as the crack winds its way around pieces of aggregate; thus a high degree of accuracy is neither practical nor necessary. The depth of cover to the rebars is also important and requires investigation by a cover-meter.

Cracking is due to many causes, but for the purpose of deciding on the need or otherwise for remedial work it can be placed in two main categories, structural cracking and non-structural cracking.

4.6.2 Structural cracking

This type of cracking would indicate that the members affected were not able to support the loads they had to carry with an adequate factor of safety.

This is discussed in Chapter 6, section 6.2.

4.6.3 Non-structural cracking

This cracking can best be described as due to any cause except the one referred to in section 4.6.2 and discussed in Chapter 6, section 6.2.

However, it must be kept in mind that in reinforced concrete, the presence of cracks can result in serious corrosion of the rebars which in turn can cause the member affected to become structurally unsafe.

The following are the main types of non-structural cracks likely to be found in an existing reinforced concrete building:

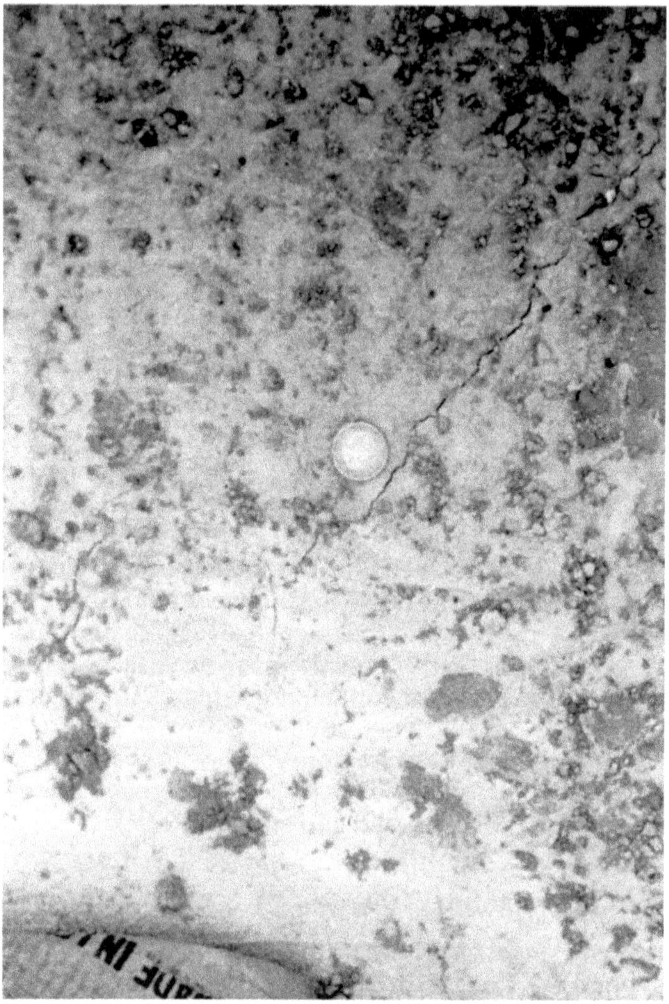

Figure 4.2 Example of plastic cracking in a floor slab.

1. plastic cracking (see Figure 4.2);
2. plastic settlement cracking (see Figure 4.3);
3. cracking caused by corrosion of reinforcement (see Figures 4.4 and 4.5);
4. drying shrinkage cracking;
5. thermal contraction cracking (see Figure 8.3);
6. map pattern cracking (crazing);
7. cracking due to bad workmanship (see Figure 4.6);
8. cracking caused by alkali-silica reaction;
9. flexural cracking.

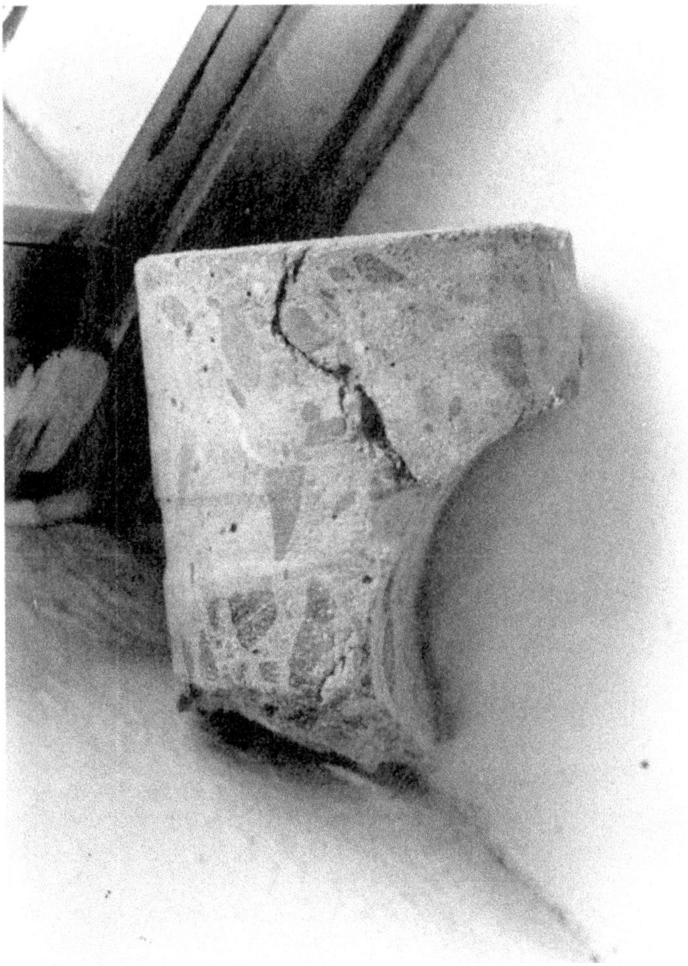

Figure 4.3 Plastic settlement cracking near the head of a mushroom-headed column.

It should be noted that Codes for structural design of reinforced concrete accept that limited cracking in the tensile zone of members is likely to occur and does no harm provided the crack width at the concrete surface is limited. BS 8110: Part 2, clause 3.2.4 suggests that maximum crack width should not exceed 0.3 mm.

This chapter is concerned with investigation of defects in existing buildings, probably more than ten years old (and more likely in excess of 25 years). Therefore only those cracks likely to be found in such buildings will be discussed here, and these are those listed under (3), (4), (5), (6), (7), (8) and (9).

Figure 4.4 Cracking and spalling in concrete column due to rebar corrosion caused by chloride-contaminated aggregates. Courtesy, A/S SCancem, Norway.

Many cracks are closely associated with the corrosion of rebars. It will be necessary to decide whether the cracks formed, due for example, to excessive tensile stress in the soffit of a beam, thus admitting moisture and carbonation to reach the steel, or whether the steel corroded (due to other causes) and the rust formation caused the concrete to crack.

4.6.3.1 Drying shrinkage cracking

This type of cracking can be caused by a number of factors, the principal ones being a badly designed mix (too much water, poorly graded fine

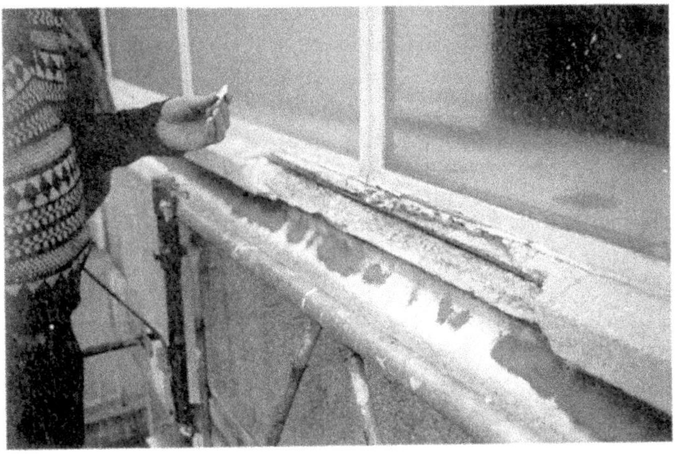

Figure 4.5 Damage to precast concrete sills caused by rebar corrosion.

aggregate containing a high proportion of very fine material), and inadequate curing. The higher the percentage of fine material in a mix, the higher will be the water demand for a given workability.

All concrete and mortar shrinks on drying out and drying shrinkage will tend to widen cracks caused by other factors. The total shrinkage (moisture movement) is made up of irreversible shrinkage and reversible shrinkage. On initial drying out an appreciable amount of the total shrinkage is irreversible, but after several cycles of wetting and drying the shrinkage becomes almost entirely reversible.

The use of shrinkable aggregates (see Chapter 1, sect. 2.6) can cause serious shrinkage in concrete made with them. As far as the UK is concerned, shrinkable aggregates are only found in the north of England and Scotland.

Shrinkage can be partly restrained and crack width and spacing controlled by the use of reinforcement.

My experience is that drying shrinkage cracks are generally confined to:

1. non-structural members which have only nominal reinforcement for handling, such as precast units;
2. floor toppings, screeds and rendering;
3. parapet walls with inadequate distribution steel; the drying shrinkage stresses often augmented by thermal contraction stresses.

4.6.3.2 *Thermal contraction cracking*

During the setting and hardening process of concrete, considerable heat is evolved by the chemical reaction between the cement and the mixing

Figure 4.6 Cracking and spalling at arris initiated by careless removal of formwork.

water. This results in an appreciable rise in temperature in the concrete. The peak temperature, the time taken to reach the peak, and then cool down, depends on a large number of factors. As the temperature of the maturing concrete rises the concrete expands and as it cools down it contracts. The coefficient of thermal expansion (and contraction) depends on a number of factors, the principal ones being the type of aggregate and the mix proportions. With similar mix proportions, a limestone aggregate concrete has a significantly lower coefficient of thermal expansion than concrete made with, say, a flint gravel.

Thermal contraction cracking is mainly due to inadequate distribution steel to take the stresses arising when the maturing concrete cools down from its maximum temperature.

It can occur in parapet walls, retaining walls, cantilever balcony slabs and sometimes in floor slabs. The 'water retaining' Code, BS 8007, deals with this design problem in some detail.

4.6.3.3 *Map-pattern cracking (crazing)*

This is really a special type of drying shrinkage cracking and can be very difficult, if not impossible, to eliminate completely. It can, and sometimes does, occur in high-quality precast units, but is more commonly found in high-quality concrete floor toppings such as granolithic and in situ terrazzo. The crack widths are usually in the range of 0.1 mm to 0.3 mm and the depth seldom exceeds 0.5 mm. This type of cracking occurs in the very early life of the concrete, but, due to the fineness of the cracks, is often not noticed/reported for months or even years. This type of cracking does not adversely affect the durability of the concrete. In precast units made with white Portland cement the cracks are liable to become filled with fine dirt and this will make the cracks more visible.

This type of cracking is sometimes mistaken for a somewhat similar pattern of cracks due to alkali–silica reaction. See section 4.6.3.5 below, and Chapter 3, section 3.6.

4.6.3.4 *Cracking due to bad workmanship*

My experience is that cracking due to bad workmanship usually arises from deficiencies in the removal of formwork, such as premature removal, lack of care in removal, and the use of the wrong type of release agent. This can result in incipient damage to the arrisses of beams, columns and floor slabs.

This damage is often repaired as part of the original contract, but the repair mortars suffer drying shrinkage and a crack develops around the perimeter of the repair. In the course of time the crack widens, allows the entry of moisture which freezes in cold weather and this widens the crack and results in spalling. If the crack extends down to the rebars (as is often the case), corrosion of the rebars occurs, which also results in spalling and, in extreme cases, the debonding of the whole of the repair mortar. See Figure 4.6.

4.6.3.5 *Cracking due to alkali–silica reaction*

The effect of alkali–silica reaction (ASR) on concrete has been described in Chapter 3, sect. 3.5.13. On flat surfaces it looks like a bad case of map cracking, as can be seen in Figure 3.6.

A clear indication that such map cracking is ASR is the exudation from the cracks of a yellowish white gel, but this can look like discoloured lime-bloom. Special care is needed in assessing the cause of map cracking in external concrete.

4.7 DIAGNOSIS OF NON-STRUCTURAL DEFECTS

4.7.1 Introduction

The definition of 'diagnosis' in the Oxford Dictionary is:

> Identification of disease by means of patient's symptoms etc. [...] formal statement of this.

For the purpose of this book, diagnosis can be translated to mean:

> Identification of the cause of the defects by careful consideration of the results of investigations and testing, and recording this in a formal report.

In Chapter 2, information was given on the main characteristics of the principal materials used for construction in reinforced concrete; Chapter 3 set out the basic causes of corrosion of steel reinforcement in concrete and the causes of physical and chemical deterioration of the concrete.

This chapter has described the investigations needed to establish the details of the defects found in existing reinforced concrete building structures.

From all this information the engineer would be expected to diagnose the reasons for the deterioration and put forward proposals for remedial work. The engineer may be asked by his client to list the causes of the deterioration in order of importance and to apportion responsibility for their occurrence. Such a request from the client could indicate that legal action was contemplated; comments on the implications of this development have been given in Chapter 1, section 1.4 to which the reader is referred.

4.8 THE ENGINEER'S REPORT TO THE CLIENT

This report should consist of:

1. a title page;
2. a contents page;
3. the instructions received from the client;
4. a description of the structure, including building records seen (if any);
5. a description of the investigations carried, including a summary of the defects found;
6. diagnosis, of the defects found based on the detailed investigation;

7. recommendations for remedial work;
8. conclusions;
9. appendixes:
   - photographs, listed, with captions;
   - photocopies of relevant extracts from Standards and Codes;
   - laboratory report;
   - drawings showing location of major defects.

The report should enable the client, in consultation with the engineer, to come to a basic decision on what he wants done. This decision can vary from doing nothing to demolition and rebuilding the structure in whole or part, depending largely on the client's long-term plans for the structure and the estimated cost of the remedial work needed.

4.9 FURTHER READING

American Society for Testing Materials (1980) *Standard Test Method for Half-cell Potentials of Reinforcing Steel in Concrete*, ANSI/ASTM C.876–80.

British Standards Institution (1991) *Concrete*, BS 5328: Parts 1–4.

British Standards Institution (1985) *The Structural Use of Concrete*, Parts 1 and 2.

British Standards Institution (1996) *Recommendations for Determination of the Initial Surface Absorption of Concrete*, BS 1881: Part 208.

British Standards Institution (1988) *Testing Concrete*, BS 1881: Part 124: Methods of Analysis of Hardened Concrete.

Building Research Estab. (1988, rev. 1991) Alkali aggregate reactions in concrete; *Digest*, 330, March, 8.

Cantor, T.R. (1984) *Review of Penetrating Radar as Applied to Non-destructive Testing of Concrete;* ACI Insitu/Non-destructive Testing of Concrete, SP-82 581–602.

Concrete Society (1992) *Non-structural Cracks in Concrete*, TR. 022.

Concrete Society (1991) *Damage Assessment, Repair Techniques and Strategies for Reinforced Concrete*, CS.108, 335.

Currie, R.T. (1986) *Carbonation Depths in Structural Quality Concrete*, Building Research Estab. Report, 19.

Sims, I. (1992) The assessment of concrete for ASR, *Concrete*, Mar./April, 42–6.

Walton, P.L. (1993) *Effects of Alkali–silica Reaction on Concrete Foundations*, BRE Information Paper IP16/93, Oct, 4.

5

Non-structural repairs to reinforced concrete

5.1 DEFINITION

A non-structural repair is one which, when completed, will not increase the existing load-carrying capacity of the member nor of the structure as a whole.

5.2 PREPARATIONS FOR REMEDIAL WORK

This work would normally be carried out as a result of the investigations and diagnosis described in Chapter 4.

It is assumed the client has accepted the engineer's report, and that this has been confirmed in writing to the engineer who would then proceed as described below. He may decide to recommend that on the completion of the work a system be installed to monitor the effectiveness of the repairs in inhibiting the further corrosion of the rebars, and this is briefly described in Chapter 6, section 6.8.

5.2.1 Contract documents

1. Except for very large and/or complex jobs, I do not favour the use of bills of quantities for concrete repair work. The use of such bills necessitates the employment of two quantity surveyors on site (one for the contractor and one for the client). Unless they are full-time on site, work will inevitably be held up pending the measurement of the areas and depths of a large number of prepared areas prior to the placing of the repair mortar.
2. The preparation of a schedule of work should list the main items of work to be carried out, but without reference to the quantity of work (number, areas etc.). The adoption of this system does mean that the contractors have to assess the amount of work actually required on each of the listed items, price them and submit the priced schedule

for the whole job. Such assessment can only be effectively done by experienced contractors. However, the tender price should reflect the significant saving in time and administration compared with the use of bills of quantities.

The schedule of work should include an item for the provision of access equipment, scaffolding and/or cradles, unless cradles are already available on the building. It may also be necessary to provide 'fans' to protect the public from debris falling from the building as the preparatory work is carried out. If so, this should be included as a separate item. The schedule should also make it clear that the cost of compliance with all contract conditions must be allowed for in the pricing of the various items of work, or as a separate lump sum.

3. The engineer would then decide on suitable general conditions of contract, based on the JCT Agreement for Minor Building Works, or the ICE Conditions of Contract for Minor Works.

5.2.2 Tendering

The engineer would send out invitations to tender to three firms known to him as being experienced in the type of work described in the contract documents, having first contacted them by telephone to ascertain if they are interested to tender for the work. The engineer should agree with the client on whether the tenders are to be returned to the client or to the engineer. I prefer the former procedure.

The issue of the tender documents is usually followed by some discussions with the tenderers to clear up any queries. One or more contractors may submit alternative proposals, often involving the use of different materials to those specified. This can create a problem for the engineer in submitting recommendations to the client for the adjudication of the contract.

Another source of difficulty in adjudication can arise when there is a wide range in the tender figures.

A careful scrutiny of the prices given in the schedules of work may show that one or more items are seriously under-priced; this may be due to an error or to deliberate policy by one contractor in the hope of being able to claim substantial 'extras'.

It is a debatable point whether the engineer should draw the tenderer's attention to such apparent error(s). It is not possible to give specific advice because a decision depends on the details of each case. One thing I wish to emphasize is that the engineer should not hesitate to recommend the acceptance of a tender which is not the lowest if he considers that the lowest tenderer will have considerable difficulty in completing the work satisfactorily and within the contract period. This problem is also referred to in Chapter 1, section 1.7.

It should be remembered that if the contractor goes into liquidation, the final cost of the work by another contractor may be double the original figure, quite apart from consequential losses and possible litigation arising from the delay. The client may not be prepared to accept the engineer's recommendation, but the engineer should make the position clear.

It is normal practice for the engineer to be responsible to the client for the inspection and control of the work including certification of contractor's payments-on-account, final account and settlement of any claims.

5.3 THE EXECUTION OF THE REPAIRS

5.3.1 Preparatory work

Each member should be carefully checked, both visually and by 'hammer testing' to ensure that all areas of defective (cracked, loose or hollow-sounding or honeycombed) concrete are clearly marked.

All defective and deteriorated concrete must be cut away, and all corroded reinforcement thus exposed must be thoroughly cleaned of all corrosion products. (See Figure 5.1.)

There is no need to cut away concrete from behind the rebars unless the rebars are significantly corroded on the inside surface, which in my experience is seldom the case. It should be remembered that there is often some rust on rebars when the concrete was originally placed, and this does no harm. (See Figure 5.2.)

This preparatory work is usually done by pneumatic tools which create considerable noise, vibration and dust. In some cases, e.g. hospitals, this is unacceptable and some other methods of removing the defective concrete and cleaning the rebars must be used.

One method is to use high-velocity water jets for cutting away the concrete and for cleaning the rebars.

The water jets leave the concrete and the steel clean and damp. A very thin film of iron oxide (rust) will quickly form on the cleaned rebars but this will do no harm; in fact it should increase bond with the repair mortar.

The pressure at the nozzle of these jets is likely to be in the range of $21\,N/mm^2$ to $69\,N/mm^2$, depending on the work to be done by the water jets.

Grit blasting may be used for cleaning rust from the rebars and final preparation of the concrete to receive the repair mortar.

In both cases care must be taken to protect adjacent parts of the building including windows, and to ensure that Health and Safety Regulations are complied with.

Figure 5.1 Close-up view of concrete surface prepared for application of repair mortar or concrete.

Whichever method is used the coarse aggregate should be exposed and all grit, dust etc. be removed from the exposed surface. As far as practical, each defective area of concrete should be cut out so as to be square or rectangular, and to a reasonably uniform depth.

Figure 5.2 Preparation of concrete beam damaged by corrosion of stirrups.

5.3.2 Grouting (bond coat)

On the completion of the preparatory work, as described in section 5.3.1, the exposed and cleaned reinforcement should be given a coat of grout consisting of Portland cement gauged with Styrene Butadiene Rubber (SBR) latex; the proportions are usually 1 to 1.5 parts cement to 1 part of latex by volume. The proportions can be varied to suit the required consistency. The prepared surface of the concrete should be damped down about one hour before the application of the grout. The grout should be well brushed into the concrete surface.

Depending on how the work is organized, the grouting of the rebars can preceed the grouting of the concrete by 24 hours or more, in which case the rebars should be given a second coat of grout at the same time as the concrete.

It is essential that the repair mortar should be applied within 20 to 30 minutes of the application of the grout, i.e. while the grout is still tacky.

To maintain the proper sequence of operations requires careful organization of the work. Unless this is achieved, bond failure between the repair mortar and the base concrete can occur.

5.3.3 The mortar mix

The recommendations which follow apply to defective areas which are suitable for repair with mortar rather than concrete. As described in section 5.3.9, the repair of honeycombed concrete may be better repaired with concrete or gunite (pneumatically applied mortar), or by pressure grouting.

There are differences of opinion on the type of repair mortar to be used; these are listed in my order of preference:

1.
   - Prebagged sand, 3 parts (by mass); the grading to comply with Grading M, Table 5, BS 882, or BS 1199 and 1200, Amend. No. 3, Table 1, Type A.
   - Ordinary Portland cement, 1 part by mass; preferably also prebagged to ensure correct mix proportions and to save time on site.
   - Gauging liquid, consisting of water and latex; the latex proportion being 10 litres to 50 kg OPC. The amount of water must be sufficient to ensure adequate workability for the method of application, e.g. by spray gun or hand application. The water/cement ratio should not exceed 0.4. The low w/c ratio will increase the rate of hardening of the mortar and reduce drying shrinkage.
2. Proprietary repair mortars, usually consisting of OPC, graded sand, prebagged for mixing on site, with a gauging liquid containing a polymer. A separate primer for application to the rebars is sometimes included. The whole package is often described by the suppliers as a 'concrete repair system'.

   Such proprietary systems are expensive, some cost up to 15 times the cost of the 'system' described in (1) above. I am not convinced that the extra cost is justified except in exceptional circumstances.

5.3.4 The application of the mortar

While the mortar for the majority of patch repairs is applied by hand, I prefer that the application should be done by spray, although this may not be practical for very small areas. However, each repair area must be finished by hand trowelling. The spray gun referred to is low pressure, compared with the 'guns' used for the application of gunite and sprayed concrete.

Whichever method is used (hand application or spray gun), it is essential for durability that the mortar be very well compacted into place and within 20–30 minutes of the application of the bond coat.

The use of the SRB latex is recommended to improve bond, to reduce the water/cement ratio while maintaining workability, and reduce the absorption and permeability of the repair mortar. For further information on polymers (which include SBRs), the reader is referred to Chapter 2, section 2.7.4.

5.3.5 Curing

I am aware that there are differences of opinion about the need to cure patch repairs, particularly if a polymer modified mortar is used. Nevertheless, I consider it desirable that all practical measures be taken to ensure proper curing of the newly placed mortar. This can present a serious practical problem on vertical surfaces, particularly when the repaired areas are small and widely spaced. In favourable circumstances, polythene sheet screens can be used, but for small isolated areas, some other solution has to be found. One option is to use water sprays but to be successful constant supervision is required. Another and more practical option is to use a spray-applied curing compound, but this also can create problems.

If, as is often the case, the whole surface of the repaired member is to be finished with a protective/decorative coating, or some other surface finish, then it is essential to obtain a curing compound that will not adversely affect the bond between the substrate and the applied finish. If there is any doubt about the effect on bond by the curing compound, then a trial patch should be arranged in advance of the start of the work.

A further point that needs attention is to ensure that the curing compound is compatible with the repair mortar if the latter is modified by the addition of a polymer, such as SBR, acrylic etc.

The function of the curing compound is to retain the moisture in the mortar until it has matured sufficiently to resist drying shrinkage stresses.

5.3.6 Finishing procedures

Due to the difficulties in ensuring proper curing, it is often found that fine hairline cracks appear around the perimeter of the repaired areas due to drying shrinkage. It is therefore recommended that a coat of cement/SBR grout be applied to the perimeter of the repaired areas. This should be done as late in the contract period as possible to allow the maximum time to elapse for drying shrinkage to take place.

I recommend that when repairs of any magnitude are carried out, the whole surface of both the repaired areas and the areas which have not needed any repair should be finished with a durable coating. A carefully selected coating will reduce water penetration and reduce, for a limited period, the ingress of carbon dioxide. In addition, the aesthetic appearance of the repaired areas will be greatly improved.

For a detailed discussion on the use and selection of coatings (barrier systems), see Chapter 7.

5.3.7 Repairing non-structural cracks with grout or mortar

Because it is not unusual to find cracks in concrete, reference to cracks and cracking has been made in Chapter 3, section 3.2.2 and in Chapter 4, sections 4.5.3 and 4.6.

The cracks considered here are those found in sound concrete and are not associated with corroded rebars and bad workmanship which have been discussed in Chapter 4 section 4.6.3.

It is of great importance to ascertain the cause of the cracking and this will automatically indicate the type of cracks and whether they are 'alive' (i.e. whether there is movement across the crack), and whether they should be considered as 'structural' or 'non-structural'.

Generally, 'non-structural' cracks such as shrinkage cracks, map-pattern cracks (crazing) do not require special repair as these cracks are usually very narrow and only extend into the concrete a few millimetres.

If treatment is considered necessary, then brushing the surface with a wire brush, followed by removal of dust and grit, and the application of thin cement/SBR grout well brushed in, would be sufficient.

'Designed' cracks in the tension zone of beams, slabs etc. (see BS 110: Part 2, clause 3.2.4) are also considered as non- structural as the width at the surface is recommended not to exceed 0.3 mm. It is very doubtful if cracks of such a width would extend down to the rebars if recommended depth of cover was provided. It is very seldom that any attempt is made to grout in such cracks.

Thermal contraction cracks (see Chapter 4, section 4.6.3.2) usually extend through the thickness of the wall or floor and can often be considered as 'dead'. Their presence does not adversely affect the structural stability of the member, but may provide access for water and other liquids to reach the reinforcement, and therefore they need repair. The details of repair will depend on whether the crack is assessed as 'dead' or whether some limited amount of cyclic movement is anticipated. Another important factor is whether the crack is accessible from both sides of the member.

Taking the 'simple' case of floor slabs and parapet walls where the cracks are considered as 'dead', the repair can be either by crack injection (as described in Chapter 6, section 6.3) or the crack can be cut out and filled with an SBR-modified cement/sand mortar. The crack should be cut down to the rebars in the form of a V, with a surface width of about 10–20 mm. All grit and dust must be removed and the surface well wetted, preferably overnight.

The mortar mix should be 1 part OPC, 2½–3 parts by volume of clean sand, graded to 'fine' in Table 5 of BS 882, with 10 litres of SBR to 50 kg cement. The mortar should be as stiff as practicable and must be well pressed into the groove. To improve bond between the mortar and the concrete, the sides of the groove and any rebars exposed by the cutting out should be given a brush coat of SBR/cement grout immediately before the mortar is applied. It is advisable for the mortar to be cured for four days, either by a spray-applied curing membrane, or by polythene sheeting (see section 5.3.5 above).

5.3.8 Repairing 'live' cracks

It can be very difficult to decide whether the crack is 'live' or not. My experience is that it is not unusual to find that cracks have been repaired on the assumption that they are 'dead', whereas subsequently it is found that some movement is in fact taking place and the repair is a failure.

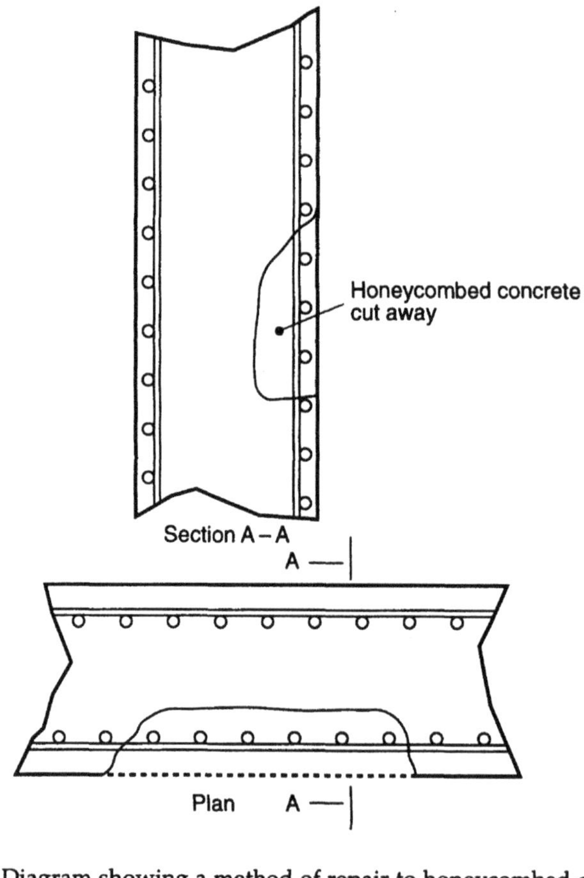

Figure 5.3 Diagram showing a method of repair to honeycombed concrete in a rc column.

The repair of cracks where it is anticipated that some degree of movement will take place across the crack requires very careful consideration, because rigid repair materials should not be used.

There is no standard way of repairing such cracks and each case has to be dealt with as an individual problem; and the reason for the movement has to be assessed before the repair method is finally determined. Attention has to be given to any existing applied finishes, such as rendering, tiling, mosaic or coatings. Unless crack injection using a specially formulated resin which possesses the desired degree of flexibility to maintain its integrity under anticipated movement, the crack will have to be cut out to a predetermined depth and width and the cut portion filled with an elastomeric compound, on either one side or on both sides if the two sides are accessible.

While I am generally not in favour of expensive proprietary repair systems, I do agree that for the satisfactory repair of 'live' cracks, proprietary materials and technical knowhow are essential.

5.3.9 Repairs to honeycombed concrete

The principal causes of honeycombing in concrete members has been discussed in Chapter 4, section 4.4.9.1. The method adopted for repair will depend on the location and extent of the honeycombing. On vertical surfaces, removal of the defective concrete and application of patch repair mortar, as detailed above, would be suitable. The same would apply to soffits of beams and slabs. However, on vertical surfaces, e.g. walls and columns, where the honeycombed concrete extends below the reinforcement, the use of 10 mm aggregate concrete may be a better solution. This repair is shown in Figure 5.3.

Further information on dealing with honeycombed concrete is given in Chapter 9, Part 1, section 9.4.5 and Part 2, section 9.9.2.

5.4 FURTHER READING

British Standards Institution (1992) *Methods of Test for Curing Compounds for Concrete*, BS 7542: 1992.

Cather, B. (1992) How to get better curing, *Concrete*, Sept./Oct., 22–5.

Concrete Society (1991) *Patch Repair of Reinforced Concrete – Subject to Reinforcement Corrosion*, TR. 038, 88.

Concrete Society (1992) *Non-structural Cracks in Concrete*, TR.022, 60.

Dennis, R. (1992) *Polymer Dispersions: Construction Materials Reference Book*, Paper 39, Butterworth-Heinemann, 12.

Parrott, L.J. (1988) Moisture profiles in drying concrete, *Advances in Cement Research*, **1**(3), July, 164–70.

Perkins, P.H. (1984) The use of SBR/cement slurry for bonding coats, *Concrete*, **18**(3), Mar. 2.

Plum, D.R. (1990) The behaviour of polymer materials in concrete repair & factors influencing selection, *The Structural Engineer*, **68**(17/4), Sept. 337–45.
Walters, D.G. (1990) Comparison of latex-modified Portland cement mortars, *ACI Materials Journal*, July/Aug., 371–7.

6

Structural repairs to reinforced concrete

PART 1 GENERAL BUILDING STRUCTURES

6.1 INTRODUCTION

Errors in design and the overloading of part of a structure which results in structural failure of the whole or part of a building are fortunately very rare. When this does occur, considerable publicity (some of which can be ill-informed) is given to the failure.

Change in the existing use of a building may result in the new imposed (live) loading exceeding that in the original design. This would require structural strengthening and it is almost certain that some concrete repair work would also be needed which should be carried out as described in Chapter 5.

Part 1 of Chapter 6 describes the special features of repairs to concrete structures which are intended to ensure that the structure is structurally safe. A structure may not be actually unsafe and in danger of partial collapse, but the investigation carried out by the engineer may show that major remedial work is urgently required if such a state of affairs is to be prevented from occurring in the near future.

Four major methods of repair are included in this chapter rather than in Chapter 5, although their use does not necessarily imply that structural failure is about to take place. These methods are:

- crack injection
- cathodic protection
- re-alkalization
- chloride removal extraction

6.1.1 Definition of failure

A failure can be considered as occurring in a component when that component can no longer be relied upon to fulfil its principal functions.

Limited deflexion in a floor which caused a certain amount of cracking/distortion in partitions could reasonably be considered as a defect but not a failure. While excessive deflexion resulting in serious damage to partitions, ceiling and floor finishes could be classed as a failure.

6.2 INVESTIGATIONS FOR STRUCTURAL DEFECTS

6.2.1 Introduction

This section considers the situation if the initial inspection/investigation detailed in Chapter 4 indicated that some parts of the structure may require strengthening. This can arise for three basic reasons:

1. serious deterioration of some of the structural members;
2. serious overloading of members (see 6.2.2 below);
3. proposed change of use involving substantial increase in floor loading.

6.2.2 Indications of structural defects

What are the likely signs of structural distress? No precise answer can be given to this question, but the following brief notes are relevant:

1. Diagonal cracks in beams and walls usually denote high shear stress and should be investigated.
2. Excessive deflexion in beams and floor slabs indicates that the members are over-loaded. This is also likely to show as cracking in the soffit at right angles to the main reinforcement (flexural cracking).
3. Bowing in columns and load-bearing walls is likely to cause cracking parallel to the main reinforcement.
4. Bowing in wall panels may be due to differential shrinkage/thermal effects between one face and the other.
5. Errors in the location, design and/or execution of movement joints, isolation joints, stress relief joints and sliding joints can result in cracking, spalling and distortion. This type of defect can be very difficult to rectify.

6.2.3 Investigation procedure

It will be seen from the previous section that visible cracking plays an important part in indicating that the structure or parts of the structure are suffering from structural distress. In other words, the members affected were unable to carry the loads imposed on them with an adequate factor of safety.

Such a state of affairs may be brought about by:

1. error(s) in design;
2. errors in construction (workmanship and/or materials);
3. actual loading significantly in excess of the design load;
4. physical damage, impact, explosion fire etc.
5. serious corrosion of reinforcement, which may be the result of many factors.

The engineer should make every effort to obtain copies of the structural calculations and as-built drawings. Unfortunately, this important information is often not available, in which case a 'structural appraisal' would be needed and this is time-consuming and expensive. Reference can be made to BRE Digest 366[1].

Assuming that adequate background information is available the general procedure is basically similar to that described in Chapter 4, but with the emphasis on obtaining information for a practical diagnosis of the structural shortcomings. Additional methods of investigation may include:

1. a UPV survey; see Chapter 4, Section 4.4.10.5 above;
2. an impulse radar survey; see section 6.2.4 below;
3. core testing for strength; see section 6.2.5 below;
4. load tests (seldom used); see section 6.2.6 below.

It should be noted that the above methods are supplementary to normal investigation techniques, and often used in combination.

6.2.4 Impulse radar survey

The author is indebted to GB Geotechnics for the information which follows.

A transducer containing the transmit and receive antennae is drawn over the surface under investigation at a constant speed. Pulses of energy are transmitted into the material and are reflected from internal surfaces and objects, e.g. changes in density, voids, reinforcing steel. The data is recorded graphically or digitally thus enabling a preliminary assessment on site, followed, if considered necessary, by detailed processing in the laboratory.

Radar responds to changes; it can identify boundaries between layers, measure thicknesses and assess voids and relative moisture content. The radar profile is effectively continuous, radio pulses are transmitted at around 50 000 pulses per second.

Transducers can be hand-held or mounted below survey vehicles, and can be operated up to 200 m from the recording station.

6.2.5 Core testing for strength

The location of the cores should be carefully selected to provide the information required and for checking the results of UPV and radar surveys.

The cores should be cut, prepared and tested in accordance with the appropriate National Standard; in the UK this is BS 1881: Part 120. Reference should also be made to BS 6089: Assessment of Concrete Strength in existing Structures, and to Concrete Society Technical Report No. 11.[2]

Misunderstandings sometimes arise over the interpretation of the test results. The actual test on the core will give the compressive strength of the concrete in the core. BS 6089 refers to the 'estimated in situ cube strength' which is defined as 'The strength of concrete at a location in a structural member estimated from indirect means and expressed in terms of specimens of cubic shape'. The Concrete Society Report refers to two types of strength; firstly: 'Estimated Potential Strength' which is defined as:

> The strength of concrete sampled from an element and tested in accordance with this procedure, such that the result is an estimate of the strength of the concrete provided for manufacture of the element, expressed as the 28 day BS.1881 cube strength, allowance being made for differences in curing, history, age, and degree of compaction between core and BS.1881 cube.

The report also provides for a correction for the influence of included steel.

When all these corrections have been made, the result is intended to give the 28-day cube strength of the concrete if cubes had been made and tested in accordance with BS 1881, at the time the member was cast. The intention is to provide an acceptable answer to the questions which arise in new construction when cubes fail. Many experienced engineers feel that with so many corrections only limited reliance can be placed on the results.

The second 'type of strength' referred to in the CS Report is 'The Estimated Actual Strength'; this is defined as:

> The strength of concrete sampled from an element and tested in accordance with this procedure, such that the result, expressed as an equivalent cube strength, is an estimate of the concrete strength as it exists at the sampling location, without correction for the effect of curing, history, age or degree of compaction.

The majority of investigations involving existing buildings are concerned with a reasonable assessment of **actual** strength as defined above, of the concrete in the load-bearing members.

6.2.6 Load tests

The testing described above in sections 6.2.3 to 6.2.5 should provide information on the general quality of the concrete and condition of the reinforcement. For the engineer to be able to predict with reasonable accuracy the load-carrying capacity of the various structural elements – beams, columns, floor slabs etc. – the following information would also be required:

1. original or, preferably, the as-built drawings of the structure;
2. similar information on any alterations made subsequently;
3. assessment of existing dead and live loads based on the present use;
4. assessment of dead and live loads which will arise from any proposed alterations.

When there is serious doubt about the value of the information available, consideration may have to be given to a load test on selected structural elements. It is accepted that design assumptions do not exactly match the as-built conditions; this is due mainly to the effects of composite action and load sharing. A load test on a beam or floor slab, if correctly carried out, will show how the element under test will react to the applied load under working conditions. During the test it is necessary to record deflexions, recovery on removal of load, and details of any crack development.

Load tests must be carried out with great care by an experienced firm with an experienced engineer on site during the test. Provision must be made to deal with any unexpected collapse. All necessary safety precautions must be observed.

Load tests are time-consuming and expensive and should only be carried out after careful consideration of practical value of the results.

REPAIR METHODS

6.3 CRACK INJECTION

6.3.1 Introduction

Deep cracks in massive concrete members and cracks which pass right through a structural member can often be repaired satisfactorily by crack injection using a selected polymer such as an epoxy resin.

When properly carried out, crack injection will significantly improve the structural strength of the member.

If corrosion or rebars and spalling of concrete has already occurred, then a satisfactory method of repair may be to remove defective concrete down to the rebars, clean off the rebars, remove all grit and dust etc., inject the crack with a suitable resin and then fill in the cut-out section of concrete with an SBR modified mortar as described in Chapter 5, section 5.3. In appropriate cases, this repair method can be combined with crack injection, e.g. to avoid cutting out on both sides of the member.

6.3.2 Essential features of crack injection

The essential crack injection consists of injecting a suitably formulated resin into the cracks. This should bond the concrete together across the crack and should form an effective seal against ingress of water or other liquids, and reduce the ingress of carbon dioxide.

Correct formulation of the resin is of vital importance; present-day polymer resins provide considerable scope for variations in the formulation so as to obtain optimum characteristics for each particular job.

The resins in general use are epoxy, polyurethane and polyester, in that order. Desirable qualities for the formulated resin include:

1. low viscosity, (to facilitate penetration into the crack);
2. formation of good bond to damp concrete;
3. suitability for injection in a wide temperature range;
4. low-curing shrinkage;

5. toughness (low modulus of elasticity combined with high yield point);
6. curing time to suit injection conditions;
7. resistance to aggressive chemicals may be required;
8. durability under service conditions.

6.3.3 The injection process

This work is highly specialized and should only be entrusted to firms with a proven record of success. Unsuccessful crack injection by one firm is likely to result in another (more experienced firm) being unable to rectify the situation; I have come across this unfortunate state of affairs on more than one occasion.

The crack injection process is carried out in the following phases:

1. preparation of the cracks;
2. location of injection points and surface sealing;
3. injection of resin;
4. removal of injection nipples (if used) and plugging the holes
5. removal of sealing strip and application of any surface treatment which may be required.

6.3.4 Preparation of the cracks prior to injection

This should consist of the removal of any dirt or loose weak material on the surface, followed by cleaning out of the crack if this is considered necessary. It is seldom that cracks less than 0.5 mm wide require cleaning unless they have been fouled by the use to which the structure has been put. Compressed air can be used for this cleaning work; solvents may be required in special cases.

6.3.5 Location of injection points and surface sealing

The distance apart of the injection points will depend largely on the depth and width of the crack. The object is to have as few injection points as possible consistent with maximum resin penetration with low operating pressure. They are either holes drilled on the line of the crack, or nipples screwed into the concrete; the use of nipples is usually reserved for high pressure work (See Figure 6.1.).

6.3.6 Injection of the resin

As stated above, crack injection work should only be entrusted to

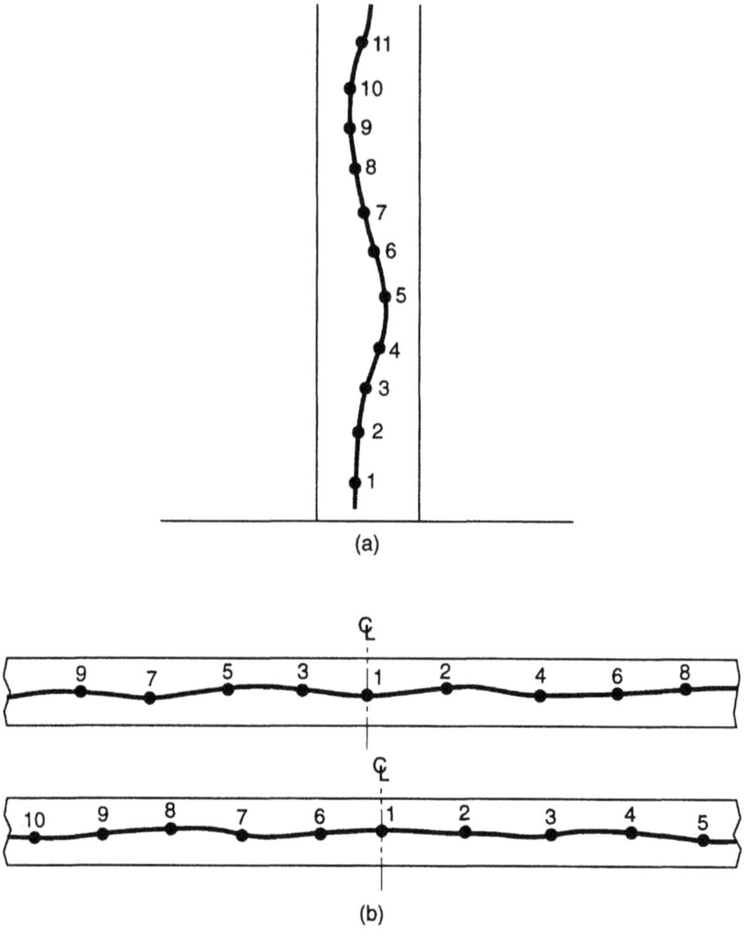

Figure 6.1 Diagram showing alternative sequences of injection points in concrete wall.

specialist firms with a good 'track record' and preferably to firms which formulate their own resins. The correct formulation of the resin to suit the particular requirements of each job is of primary importance. If the formulator and the applicator are the same firm, any question of divided responsibility is avoided. For larger jobs, some adjustments in formulation may be required during the execution of the work to meet unforeseen conditions and this requires close co-operation between the staff on site and the laboratory.

Firms specializing in crack injection develop their own techniques and equipment. Some firms favour simple means of injection by gravity feed or pressure guns, the resin being premixed in batches. Others use sophisticated equipment for continuous feed of freshly mixed resin and hardener through separate feed pipes, bringing the two materials together at a specially designed nozzle. One firm supplements pressure-feed by the use of a vacuum mat to assist penetration.

In appropriate cases, the degree of penetration can be checked by coring or by a UPV survey or impulse radar.

The aim of all injection processes is to obtain uniform penetration of the resin and complete filling of the cracks. It has been found in practice that a deliberate fluctuation in the injection pressure can be more effective than an increase in continuous pressure. The volume of resin used in filling cracks is very small.

Where cracks are inclined or vertical, it is usual to commence injection at the lowest injection point and work upwards. For horizontal cracks there is no fixed order of work; the injection can start at one end and work along the crack to the other end, or start in the middle and work first left and then right, or alternately left and right.

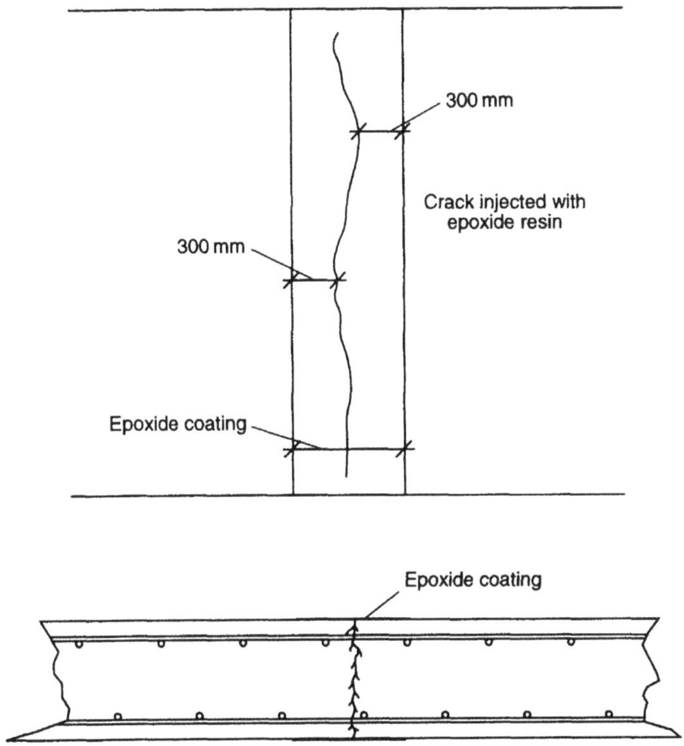

Figure 6.2 Diagram of crack injection and coating of thermal contraction crack in concrete wall.

6.3.7 Final work following injection

It is usual to remove the injection nipples (if they are used) and seal the holes when the resin has set.

Crack injection, except where the cracks are very wide on the surface, is likely to be much less conspicuous than cutting out and repairing with mortar. Nevertheless, the injected crack will be visible. Where appearance is important, surface grinding and some 'cosmetic' treatment will help to mask the crack.

Figure 6.2 shows a repair using crack injection.

6.4 REPAIRS USING EPOXY-BONDED STEEL PLATES

6.4.1 Introduction

It appears that the use of steel plates bonded to the tension face of reinforced concrete beams was developed in France and South Africa in the 1960s. In the UK the first recorded use of resin-bonded steel plates to strengthen an existing building was in about 1966; for the strengthening of road bridges the first use was in 1975 for the Quinton Interchange on the M5.

The use of steel plates bonded to the sides of beams to increase shear resistance would appear to be a possibility, but I have not found any record of such use. However, it has been used to stiffen reinforced concrete floor slabs by bonding the steel plates to the top surface of the slabs, and to strengthen the connections between beams and columns.

6.4.2 Information on the technique

This technique depends on composite action between the steel plates and the concrete, and therefore requires maximum bond between steel-epoxy resin-concrete. This in turn requires very careful preparation of the contact surfaces. The steel must also be adequately protected against corrosion on the exposed surfaces. When the work is properly carried out and the bond strength has fully developed, the strength of the epoxy joint should exceed that of the concrete which will fail in horizontal shear.

The strength of epoxy resins deteriorates rather rapidly at temperatures in excess of about 65°C and this constitutes a fire hazard when used in building structures, but is of little significance in bridges.

A considerable amount of work has been carried out on the development of this technique at the Transport and Road Research Laboratory at Crowthorne, and at Sheffield University, and the universities of Dundee and Warwick.

The technique is discussed in detail in the Report by the Standing Committee on Structural Safety, set up by the Institution of Civil Engineers and the Institution of Structural Engineers. The Report recommends caution in the use of this method and emphasizes that it should not be used to strengthen structures in which the concrete is poor quality, or the reinforcement has suffered chloride attack, or the concrete has been damaged by alkali–silica reaction, unless these basic defects are first corrected. Two methods of dealing with chlorides in the concrete which have corroded the rebars are discussed in sections 6.6 and 6.7 of this chapter.

6.5 THE USE OF FIBRE-REINFORCED PLASTICS

According to *Research Focus*, No. 21, April 1995, the use of fibre-reinforced plastic (FRP) as a substitute for steel has been investigated in Switzerland, Germany and the USA, and at Oxford Brookes University under an EPSRC (Engineering and Physical Sciences Research Council) grant. I have not come across a report on the use of these materials for repair work.

6.5.1 Carbon fibre composites

Reference has been made in section 6.4 to strengthening reinforced concrete beams by epoxy-bonded steel plates. An alternative to this method by using carbon-fibre reinforced composites was reported by Dave Parker.[3] It is stated in the article that to provide the additional strength required, the epoxy-bonded steel plates would have had to be 7 mm thick, but these were replaced by a 1 mm thickness of carbon fibre composite (a Sika Carbodur System). The work was part of a major refurbishment.

6.6 CATHODIC PROTECTION OF REINFORCEMENT

6.6.1 Introduction

If an investigation shows that the principal reason for the corrosion of reinforcement is the presence of chlorides in the concrete, then normal repair methods will not be durable, resulting in the need for a continuous programme of repair.

This leads to the question of what can be done to ensure a satisfactory and durable repair. The main method to deal with this problem is the use of cathodic protection.

6.6.2 General principles of cathodic protection

Cathodic protection has been used for many years for the protection of steel and iron structures such as jetties, storage tanks, and underground pipelines.

In the early 1960s highway engineers in the USA and Canada started trials on the use of cathodic protection for the reinforcement in bridge decks which had been severely damaged by chloride attack from deicing salts. It is now used extensively in both USA and Canada for this purpose, mainly the impressed current system. In the UK cathodic protection is being increasingly specified to combat chloride corrosion of rebars, mainly in highway bridges.

Corrosion of steel is an electro-chemical process and exposed steel in a moist environment will corrode due to differences in electrical potential on the surface of the metal itself. These areas form anodes and cathodes, and this allows an electric current to flow from anode to cathode; the metal suffers corrosion at the anodic areas. The objective in cathodic protection is to ensure that the steel to be protected forms the cathode.

There are two practical ways of introducing cathodic protection to steel:

1. by connecting the steel to a metal which is 'less noble' in the electrochemical series (this forms the anode); this is known as sacrificial anode protection;
2. by the application of an external electric current of sufficient intensity to 'swamp' the corrosion current; this is the impressed current technique.

1. Sacrificial anodes

The following list shows part of a basic series in which the metals high on the list become anodic to those lower down and hence provide protection. This can be described as sacrificial anode protection and depends on one metal being designed to corrode and so prevent another metal connected to it from corroding.

- sodium
- magnesium
- zinc
- aluminium
- mild steel
- cast iron
- stainless steel (cromium-based)

When two dissimilar metals are in contact in an electrolyte, a current is produced as a function of the electrochemical series of metals. For example, if steel is in electrical contact with zinc in an electrolyte, a current will flow from the zinc to the steel because zinc is anodic to steel and the zinc will corrode but the steel will not. This is the principle of the use of sacrificial anodes.

2. *Impressed current*

Protection can be provided by the introduction of an impressed current (dc).

The structure to be protected is connected to the negative supply (the cathode) and the positive to an introduced anode which is specially selected to have semi-inert or non-corrodable properties. The range of suitable anodes is limited and is designed to last very much longer than the 'sacrificial' anodes referred to in the previous paragraph, and a useful life of 20 years or more can be expected. The materials available are graphite, platinized titanium, high-silicon iron, tantalum or niobium; lead/silver alloys are often used in marine installations.

The relevant British Code of Practice is BS 7361: Cathodic Protection; the USA Code is ASTM B.843

In considering cathodic protection to an existing structure the following matters need careful attention:

1. The type of protection which will be the most suitable in the given circumstances, i.e. sacrificial anodes or impressed current.
2. The use of cathodic protection to reinforcement encased in concrete introduces special problems. The pore water in the concrete acts as the electrolyte of the corrosion cells.
3. The impressed current must be adequate to suppress the corrosion current. Concrete has a high resistance to the flow of electric current and this has to be overcome.
4. It has been claimed that cathodic protection can cause hydrogen embrittlement but I have not seen a reliable report of such a case.
5. If cathodic protection by means of impressed current is used in a building, care must be taken to ensure that corrosion of unprotected ferrous metals in the building does not occur.

The use of a cathodic protection system for a reinforced concrete structure requires special knowledge and experience in concrete technology in addition to that required for 'normal' cp systems used for steel structures. For the system to be successful in a reinforced concrete structure a number of factors should be taken into consideration, the principal two being:

1. the possibility of initiating alkali–silica reaction;
2. the existence of significant discontinuities in the reinforcing steel.

6.7 RE-ALKALIZATION AND CHLORIDE EXTRACTION

6.7.1 Introduction

In the early 1990s methods for the re-alkalization (R–A) and chloride extraction/removal to remedy corrosion of reinforcement were introduced into the UK from Norway where it had been developed and patented by Norwegian Concrete Technologies.

Generally, re-alkalization may be appropriate in cases where the corrosion of the rebars has resulted from carbonation of the concrete in contact with the steel. For information on carbonation of concrete see Chapter 3, section 3.2.5.

When the corrosion of rebars is caused by the presence of chloride ions in the concrete, the use of chloride removal may be appropriate. For information on chloride induced corrosion of rebars see Chapter 3, section 3.2.6.

Both methods are electro-chemical in character.

6.7.2 Re-alkalization

This supplements normal repair (patch repairs) to defective concrete and is not an alternative method of repair. It restores the alkalinity of the concrete in contact with the rebars, so that the pH is above 10.0 which should be sufficient to restore the passivation of the steel. As will be seen from Chapter 5, patch repairs to concrete, if properly carried out, will restore the alkaline environment around the rebars in the damaged areas.

The process of R–A is briefly described as follows.

An anode, usually cellulose, is fixed to the surface of the concrete which is kept moist by a solution of sodium carbonate which acts as an electrolyte. The alkaline solution penetrates into the concrete by electro-osmosis; electrolysis at the surface of the rebars produces hydroxyl ions, thus creating a highly alkaline environment around the steel.

The process needs experienced operation and control. The increase in alkalinity of the concrete can have unfortunate side effects if the concrete has been made with silicious aggregates (see Chapter 3, section 3.5.13 – Alkali–silica reaction).

6.7.3 Chloride extraction/removal

This technique can be considered as an alternative to cathodic protection in appropriate circumstances. Chloride-induced corrosion of steel reinforcement has been described in Chapter 3, section 3.2.6.

Chloride ions are negatively charged and so an electro-chemical process can be used to remove the chloride ions in the concrete in contact with the steel and move it to an external anode. The anodes used can be those used for cathodic protection, and as previously stated the principal anode in current use is coated titanium mesh. This is fixed temporarily to the outside of the concrete member to be treated and is embedded in material which is kept wet during the chloride extraction; an alternative method of providing a wet surface can be used.

It is not practical to remove **all** chloride from the concrete, but if properly carried out the chloride content of the concrete cover to the rebars can be reduced to an acceptable level. If there has been a massive penetration of chloride into the concrete beyond the rebars, then resistance to the passage of the current increases and the removal of the chloride becomes increasingly difficult. It is therefore important to ascertain, by sampling the concrete, the variation of chloride concentration with depth from the surface before a final decision is taken to use this system.

The system does not deal with the re-entry of chloride ions into the concrete (from say, de-icing salts), which would result in corrosion starting afresh. Under these circumstances, it cannot be compared with cathodic protection which provides continuous protection to the rebars irrespective of the ingress of chloride ions.

Two known side effects which can arise from chloride removal are:

1. initiation or acceleration of alkali silica reactivity, as the reaction at the surface of the rebars may raise the alkalinity of the concrete to a dangerous level when the aggregate used is siliceous;
2. reduction of bond between rebars and concrete. This will depend largely on the type of bar surface, i.e. ribbed or smooth. In most cases this is not a problem, but should be kept in mind if smooth surface bars are used.

6.8 MONITORING CORROSION OF REBARS AFTER COMPLETION OF REPAIRS

The loss of section (caused by corrosion) of the rebars can be determined by measuring the electrical resistance of a probe. The probe must be made from the same type of steel as the rebars and must be of small sectional area, i.e. thin, so that a limited amount of corrosion will result in a significant loss of metal (reduction of cross-sectional area). Changes in temperature can cause changes in the resistance of the probe and this may exceed the changes due to corrosion. This effect must be compensated for by including a reference element which is exposed to the same temperature changes as the probe but is protected from corrosion by a suitable coating.

Other factors affecting the result are changes in moisture content of the concrete and changes in the salt content. Each reading shows the total loss of metal (due to corrosion) to date and from this the rate of metal loss can be calculated. The method should only be used by specialists in this field as interpretation is difficult.

Another method, developed by Colebrand Ltd, is to embed a half-cell (copper-copper sulphate or silver-silver chloride) in the member to be monitored, linked to a data logger which can take readings automatically at intervals and store the results for collection and analysis.

The time factor must be borne in mind as this may run into years.

PART 2: REPAIRS TO FIRE-DAMAGED CONCRETE BUILDINGS

6.9 INTRODUCTION

The effect of fire on structural elements of a building depends on the nature of the materials used and their resistance to thermal damage, the temperature reached and the duration of the fire. Concrete is incombustible and very resistant to the effects of fire. An important British Standard is BS 476, Fire Tests on Building Materials and Structures, which is in some 18 parts. Other legal requirements relating to minimum cover to reinforcement are contained in BS 8110 – Parts 1 and 2, Structural Use of Concrete.

A fire-damaged structure should be inspected as soon as practical after the fire. The reason for speed is mainly economic because a building which has experienced a fire of any significance is unlikely to be useable until the damage has been assessed and decisions taken on necessary remedial work.

It may be possible to restore part of the structure quickly, e.g. part(s) which have suffered only water damage, while other parts may require detailed assessment as described below.

For concrete structures the options are usually restoration or demolition and rebuilding, or some combination of these.

The damage has to be assessed in relation to the options mentioned above. A practical approach is to consider the problem under the following headings:

1. the age and existing use of the building, and whether any change in use is contemplated;
2. the percentage of the building which has been damaged;

3. the fire intensity in various parts of the structure;
4. the structural significance of the badly damaged parts or components (columns, beams, slabs etc.);
5. whether a detailed investigation is likely to be necessary or justified, as such an investigation may include extensive testing and is both expensive and time-consuming.

6.10 PRELIMINARY ASSESSMENT

It is essential first to ensure that there is safe entry to those parts of the building that require inspection.

Initially the inspection will be a visual one probably combined with simple removal of small pieces of loose/spalled concrete to provide a better view of colour changes (if any) in the concrete and condition of the exposed reinforcement.

Time spent on this examination and the clear recording of conditions found are fundamental to a reasonable assessment of the remedial work required.

Special features to note include:

1. extensive cracking and spalling and distortion/deflexion of members in areas of structural importance;
2. existing use of the building and its contents, the latter are of particular importance as they may have had a significant effect on the damage done. For example, PVC is inflammable and when it burns chlorine compounds are generated which in the presence of moisture (of which the fire-fighters ensure there is plenty available), forms hydrochloric acid which vigorously attacks both steel and concrete. The effect will of course depend on the quantity of PVC which has been burnt and the amount and concentration of the acid formed.

6.11 THE EFFECT OF FIRE ON CONCRETE AND STEEL REINFORCEMENT

For the purpose of this discussion, 'concrete' means Portland cement concrete made with aggregates from natural sources.

The effects of fire on concrete are complex and depend on a large number of factors, including type of aggregate, the design of the structure and its ability to accommodate movement arising from large changes in temperature, and the temperature reached. Colour changes in concrete can be a useful indicator of the temperature reached by the concrete.

A pink colour generally indicates that the concrete affected by the colour change has reached a minimum temperature of about 300°C, but this temperature may have been reached without this resulting in any colour change. Therefore reliance should not be placed on the absence of a colour change for assessment of the temperature reached by the concrete and consequent loss of strength. Concrete which shows this pink discoloration has suffered a significant and irreversible loss of strength. If the pink colour extends down to the rebars, then an investigation into the strength of the concrete between and beyond the rebars is necessary. It is also desirable to check the strength of the steel. The temperature reached by the concrete will reduce as distance from the surface is increased, except where spalling has occurred during the fire.

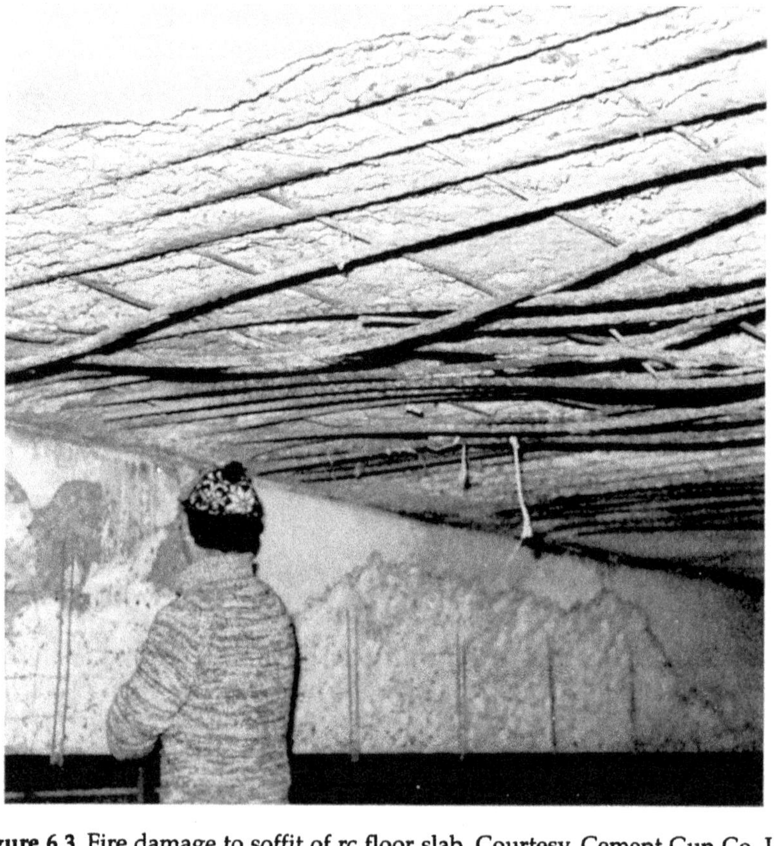

Figure 6.3 Fire damage to soffit of rc floor slab. Courtesy, Cement Gun Co. UK Ltd.

Steel rebars which are buckled or distorted are likely to have been exposed to fire. A temperature of about 600°C will result in a loss of strength up to 50%. This relation between buckling and temperature applies particularly to rebars in beams and slabs, but not necessarily to columns as the rebars are much more restrained due to the use of ties and helical binding. This indicates that exposed rebars in columns need careful consideration. (See Figure 6.3.)

Prestressing wire loses about 50% of its strength at a temperature of around 450°C.

Lightweight aggregate concrete is more resistant to fire damage than concrete made with natural aggregates such as flint gravels, granites and limestones. This is due to a number of factors, in particular its thermal insulating properties. This reduces the temperature gradient with depth from the surface and also reduces the effect of thermal shock when cold water from fire hoses is discharged onto the hot concrete.

6.12 ON-SITE TESTING

While the importance of a careful and informed visual examination of the fire-damaged structure cannot be over-emphasized, some on-site testing is usually required. This can vary from simple 'hammer' testing to time-consuming and costly load testing.

Non-destructive testing could include ultra-sonic pulse velocity survey, an impulse radar survey, the use of the rebound hammer, or an electro-magnetic cover meter survey to locate reinforcement and determine depth of cover. These methods are very useful but the results are indirect and inferential and more direct and comprehensive information may be needed.

It may be decided to take cores and measure compressive strength of the concrete in specific locations. A careful examination of cores can provide information on the depth of colour change and deteriorated surface layers where no colour change has occurred and the depth of cracks.

The testing of steel rebars is relatively straightforward, and is carried out to determine whether the temperature reached had adversely affected the yield point and ductility.

The test results should enable the engineer accurately to assess the extent of the fire damage and from this to determine the most appropriate methods of renovation/repair for the structure as a whole, and whether the whole or part is in such a condition that demolition and rebuilding is justified.

Great caution must be used when dealing with a fire in a prestressed concrete structure, and this warning applies particularly to taking samples of concrete and steel.

The effect of fire on prestressing steel is rather different to that on standard steel reinforcement. Collapse can occur much more suddenly and with little prior warning such as would be the case with a reinforced concrete structure.

6.13 EXECUTION OF REPAIRS

One of the first considerations is to determine whether special safety requirements are needed due to the condition of the structure as a whole or parts of the structure. This is now of prime importance to the engineer in view of the requirements of the Construction (Design and Management) Regulations 1994; all the provisions of these Regulations were in force by 31 December 1995. Some general information on these Regulations has been given in Chapter 1. 'Construction Work' includes 'demolition ... construction ... alteration, renovation.' The following persons are specifically referred to in the Regulations: The Client; Planning Supervisor; Designer; Principal Contractor; Other Contractors.

As far as the repair work is concerned, in England, the requirements of the Building Regulations 1991 with regard to Structure, Internal Fire Spread (surfaces), Internal Fire Spread (structure), and External Fire Spread, must be met. The more detailed requirements of the Regulations are contained in approved Document B. Section B2, Internal Fire Spread (linings); B3, Internal Fire Spread (Structure); and B4, External Fire Spread. These documents require careful study. Needless to say, normal design practices, and the use of cement-based mortars and concrete will normally ensure compliance with the Regulations.

6.13.1 Repair materials and Execution of work

Once a final assessment of the damage has been made, decisions can be taken on:

1. parts of the structure to be demolished and rebuilt (if any);
2. members which require strengthening (if any);
3. schedule of general repairs.

Contract documents can then be prepared, and tenders invited, either on a competitive basis using a selected list, or by negotiated tender.

General repairs will follow the same principles as set out in Chapter 5 for non-structural repairs. However, due to the nature of the damage, it is likely that considerable use will be made of pneumatically applied concrete and mortar, (gunite/shotcrete). One of the main problems with this method of repair is the difficulty in ensuring that the rebars, both existing and new, are completely embedded, i.e. the elimination of voids behind the rebars. This work may be a sub-contract to the main contract and it is important to ensure that only special contractors, with considerable experience and a proven record of satisfactory work, are employed. (See Figure 6.4.)

Figure 6.4 View of floor soffit in Figure 6.3 after fixing of new reinforcement prior to guniting. Courtesy, Cement Gun Co. UK Ltd.

It is also of great importance to ensure, as far as practical, that the minimum cover of well-compacted mortar/concrete, as recommended in the relevant Regulations (in the UK, BS 8110 – Structural use of Concrete) is provided. This can prove to be very difficult in cases where the existing cover itself is inadequate, e.g. only 15 mm instead of 30 mm.

The Code (BS 8110), Table 3.5 lays down 'nominal cover to all reinforcement, including links, to meet specified periods of fire resistance'. The Code also lays down in Fig. 3.2 minimum dimensions of reinforced concrete members for fire resistance. Clearly, in difficult cases, some compromise has to be reached and early discussion with the local building control officer is advisable. It may be necessary to apply a coat of gunite over the whole of the member under repair to ensure adequate cover throughout, and to obtain line and level and surface appearance.

Patch repairs with low-pressure applied cement-based mortar (preferably modified with SBR to improve bond and reduce shrinkage cracking) are acceptable. Hand-applied mortar may be used for very small areas.

The use of epoxy resins for repair of deep, wide cracks needs careful consideration as epoxies being organic are not fire-resistant and have a low melting point (below 100°C). This means that cement-based mortar would have to be used, but the risk of shrinkage cracks must be accepted.

The final application of a durable, protective and attractive coating is desirable. For a detailed discussion on the use of coatings, see Chapter 7.

PART 3 REPAIRS TO CONCRETE HIGHWAY BRIDGES

6.14 INTRODUCTION

While this part of Chapter 6 deals briefly with **repairs** to concrete highway bridges, reference should be made to BS 5400: Steel, Concrete and Composite Bridges. This Standard is published in ten Parts, of which Parts 1, 2, 4, 7, 8, 9 and 10 are relevant to concrete bridges.

The repair of a concrete road bridge is an important undertaking and is normally only entrusted to experienced Consulting Engineers and Civil Engineering Contractors.

The basic principles which should be adopted for the repair of reinforced concrete bridges are essentially the same as those for other land-based reinforced concrete structures.

Major repairs may involve closing the whole or part of the bridge to traffic.

Such work calls for investigation, diagnosis and preparation of the usual contract documents.

For large jobs, a bill of quantities will be required.

BS 5400, Part 1, Section 6, states: 'A design life of 120 years has been assumed throughout BS 5400 (unless otherwise stated).' While this clearly applies to the design and construction of new bridges, it does place considerable responsibility on those responsible for major repair work to ensure as far as possible, very long-term durability of the repairs. It also indicates that a regular inspection and maintenance programme is essential.

While the chances of a road bridge being originally under-designed structurally in the UK is very small, *BRE Research Focus*, No. 21, April 1995 states:

> Over 10,000 of the 142,000 road bridges in the UK need to be replaced or strengthened to meet with the new 40 tonne vehicle EC Directives by 1999.

This is no reflection on the original design of the bridges as the strengthening is needed to carry much heavier loading than was contemplated at the time of the design.

The repairs dealt with in this chapter are those needed to rectify deterioration, the causes of which are discussed below, and may involve strengthening.

6.15 INVESTIGATIONS

In the UK the Department of Transport (DoT) is involved in the design, construction and maintenance of highway bridges. A number of DoT publications relevant to this chapter are included in 'Further reading' at the end of this chapter.

The DoT place great emphasis on inspections and investigations and their publications give detailed directions on how these should be carried out. Careful study of the DoT directions is essential for anyone investigating deterioration of a concrete highway bridge.

It is important when carrying out a design check to compare the results with the Codes valid at the time the bridge was constructed, because it is probable that in the mean time certain aspects of design have been tightened up. The result of the design check may indicate that the bridge is now 'under-designed', this does not necessarily mean that damage has occurred and resulted in a potential failure.

The investigation programme would normally consist of a preliminary 'viewing' of the structure at which the major faults would be noted. From this, proposals for a detailed investigation would be drawn up. This may have to include for complete or partial closure of the bridge for a limited period. Where this is an overbridge, the provision of the necessary access may interfere with the traffic at the lower level.

Figure 6.5 Impulse radar survey on Papendrecht Bridge, Netherlands. Courtesy, GB Geotechnics.

It is assumed that the engineer entrusted with the investigation is experienced in this type of work and, therefore, from his preliminary inspection he knows in basic terms what he has to look for in detail, and this is summarized below in general terms:

1. the general condition of the concrete, in terms of cracking, spalling and rust stains, evidence of water staining, excessive weathering of the concrete surface; impact damage;

2. location and classification of cracks;
3. signs of alkali–aggregate reaction (map cracking and extrusion of gel);
4. condition and adequacy of surface water drainage system;
5. types and condition of joints;
6. types and condition of bridge bearings;
7. if the bridge contains post-tensioned members, then consideration should be given to an investigation of the grouting in the prestressing ducts, with particular attention to the condition of anchor blocks;
8. cover meter survey;
9. half-cell potential survey to determine probable extent of corrosion of the rebars;
10. tests/examination to determine type of corrosion of rebars, i.e. general corrosion or localized corrosion; the latter may result in pitting corrosion;
11. UPV survey and/or impulse radar survey to locate and determine extent of honeycombed concrete, delamination, cracks.

Figure 6.6 Impulse radar survey on Khimkinski Bridge, Moscow. Courtesy, GB Geotechnics.

The use of impulse radar surveys to investigate conditions below the surface of bridge decks has increased in recent years, and I am indebted to GB Geotechnics for the following examples:

Papendrecht Bridge, Netherlands: a Radar survey was carried out to locate and map the position and depth of transverse post tensioning ducts in the bridge deck. Work was to be carried out for the construction of new crash barriers and there was concern that the drilling for the barrier fixings may damage the ducts. Radar scans were profiled at each drilling location and duct positions were marked immediately onto the surface of the deck. Fixing positions were then located accordingly, thus increasing progress by the contractor and reducing interference with traffic flow. (See Figure 6.5.)

Khimkinski Bridge, Moscow: The Radar survey was required in connection with an overall assessment of the structural integrity of the bridge. The survey provided an assessment of the bond between the surfacing and the deck slab, the location and condition of the reinforcement, and condition of the concrete with special reference to transverse joints. The survey also provided a valuable comparison between the actual as-built arrangements and the record drawings. (See Figure 6.6.)

6.16 SAMPLING AND TESTING

The number, location and type of samples should be representative and adequate for the purpose of diagnosis.

Samples would be needed for tests (1) to (9) in the following list.

1. Tests to determine depth of carbonation.
2. Cement content of the concrete; determined initially from random pieces of concrete, and more accurately from cores after compression tests.
3. Determination of cement type – if required.
4. Chloride content of the concrete and assessment of its origin (in the original concrete mix, or from external source(s) such as surface water contaminated with de-icing salts). This may point to defects in the drainage system.
5. Sulphate content of the concrete (bearing in mind that Portland cement contains about 3.5% sulphate as SO_3).
6. Assessment of void content and degree of compaction.

Additional tests may be considered desirable, the more usual of these are:

7. water absorption of the concrete (BS 1881: Part 122);
8. original water content of the concrete (accuracy doubtful) (BS 1881: Part 124);
9. aggregate grading and type (BS 1881: Part 124).

6.17 CRACKS AND CRACKING

Cracks and cracking have been discussed at some length in Chapter 4, Section 4.6 and in Part 1 of this chapter.

The cracks can be divided into two main categories:

1. structural;
2. non-structural.

To this must be added that cracks of any type can have structural implications if they are excessively wide and deep, and/or are associated with corrosion of reinforcement.

6.17.1 Causes of cracking

The most likely causes of cracking sufficiently severe to affect adversely the long-term durability of the structure are:

1. corrosion of the reinforcement (general and/or local);
2. corrosion of prestressing cables in ducts in post-tensioned members, and stress cracking at anchorages;
3. defects in the bridge deck membrane, expansion joints and construction joints, resulting in corrosion (1) and (2);
4. failure or partial failure of movement joints and bridge bearings, resulting in excessive concentration of stress.
5. aggregate–alkali reaction (aggregate–silica reaction);
6. overload.

Foundation failure is likely to result in serious cracking, but fortunately this is rare.

6.17.2 Corrosion of rebars

This can result from a variety of causes.

1. from inadequate concrete cover;
2. the presence of chlorides in the concrete derived principally from deicing salts from splash and/or from defects in the deck membrane, and defective joints. In the older bridges, constructed prior to about 1965, calcium chloride-based accelerators may have been used in the concrete mix.

6.17.3 Corrosion of prestressing cables

While this type of corrosion causes serious concern, the actual number of reported cases is relatively small. However, in September 1992, the

Department of Transport issued a temporary ban on the commissioning of new bridges of the 'grout-duct post-tensioned type.'[4]

An investigation by the Transport Research Laboratory in 1981 into the grouting of 12 post-tensioned concrete bridges constructed between 1958 and 1977 showed that voids were found in the ducts of ten bridges.

A review of published papers and other literature on the risk of corrosion of the steel tendons indicates that in the past insufficient attention was paid to the absolute need to take all practical precautions to prevent the ingress of moisture and contaminants (mainly chlorides) into the ducts.

The sudden collapse of the Ynys-y-Gwas bridge in 1985 was the third collapse of a post-tensioned concrete bridge in the UK and this caused great concern.

Since about 1992, a great deal of work has been put in hand to rectify this state of affairs. See 'Further reading' at the end of this chapter.

On 25 September 1996, the Highways Agency announced the removal of a four-year ban on grouted construction, thus again permitting the use of grouted post-tensioned concrete for highway bridges.

6.17.4 Defects in deck membrane and joints

Defects in the deck membrane have more widespread effect on initiating corrosion of rebars and prestressing cables than defects in joints.

Many road bridges constructed in the UK prior to the 1960s were not provided with a waterproof membrane for the deck. The absence of the membrane and the increasing use of deicing salts resulted in the serious deterioration of the reinforced concrete deck.

Unfortunately, the introduction of waterproof membranes as a standard requirement did not eliminate corrosion problems in bridge decks. In practice, it proved very difficult to ensure a completely waterproof membrane system with long-term durability. This was partly due to the very arduous conditions under which the membrane had to operate and partly to defects in the detailing at joints and kerb lines.

All joints should be watertight as leakage at joints allows surface water contaminated with de-icing salts to penetrate the concrete deck and also the concrete sub-structure.

6.17.5 Failure or partial failure of movement joints and bearings

When this occurs it can lead to cracking and spalling arising from restriction of movement. Therefore a careful examination of all movement joints (expansion, contraction and sliding) and bridge bearings is essential.

Movement joints should accommodate expansion, contraction, vibration and deflexion and general movements of the bridge structure arising from small but inevitable foundation movements. Joints in the deck and other members must also be watertight. Because watertightness is probably the most difficult requirement to maintain, provision should be made to drain away water that passes through the joint.

6.18 ALKALI–SILICA REACTION

Road bridges have not proved exempt from problems of alkali–aggregate reaction. In the UK this has so far been confined to alkali–silica reaction-ASR, which has been discussed in Chapter 3, section 3.5.13 and Chapter 4 section 4.6.3.5.

The Department of Transport publication BA. 52/94 (part of the Design Manual for Roads and Bridges), states:

> In recent years a considerable number of bridges in the UK have been diagnosed as suffering from ASR.

ASR arises from the reaction between the alkaline pore water in Portland cement concrete and certain types of siliceous aggregates. The reaction produces a gel which absorbs water and the resulting expansion can crack the concrete. Figure 3.6 in Chapter 3 shows map cracking typical of ASR attack.

However, it is generally agreed that for ASR to occur, the following conditions must all be present:

- high alkalinity in the concrete;
- sufficient amount of reactive silica;
- sufficient amount of moisture.

A great deal has been written about this type of deterioration in concrete structures and a number of different opinions have been expressed about the actual reduction, if any, in the load-carrying capacity of structures affected by ASR. I have not seen any authoritative reports on load tests on structures affected by ASR which show that the members tested have suffered a reduction in strength sufficient to reduce the factor of safety to an unacceptable level.

The Department of Transport in their publication BA. 52/94 makes the following comment:

> Recent research has shown that ASR has much less effect on strength than would be imagined from the appearance of the

affected members; indeed some tests have identified some increase in strength due to ASR.

A joint study by EPSRC and TRL confirms the above and the study considered the effect of restraint on expansion by the presence of reinforcement and applied external tensile stress.

6.19 REPAIRS

6.19.1 Patch repairs

Cracking and spalling – usually accompanied by rust staining in exposed structural members, such as beams, columns, parapets – which has been diagnosed as arising from poor-quality concrete cover and/or inadequate depth of cover can be repaired by standard procedures described in Chapters 5 and 6. The whole of the members which have been repaired should receive a barrier coat of selected material (see Chapter 7 for information on coating materials). It is advisable for those members which have not needed repair to be given at least two coats of the same selected coating, or two spray coats of a Silane approved by the DoT (see DoT specification BD/43/90.)

6.19.2 Cathodic protection

However, if the cause has been diagnosed as due to the presence of chlorides in the concrete, and assuming that structural integrity has not been impaired by reduction in the sectional area of rebars, then the most satisfactory long-term solution would be cathodic protection. Such a system, if properly designed and installed, would stop rebar corrosion and prevent its restarting. Careful monitoring of the installation is necessary. In a paper by Way and Virmani[5], they state:

> At the present time, the only method we believe is practical to arrest corrosion of reinforcing steel and hence, effect a permanent repair, is the application of cathodic protection.

6.19.3 Joints and bearings

Apart from cleaning out movement joints as far as this is possible and checking bridge bearings for damage and seizing-up, very little can be done to effect repairs. The answer is usually to replace the joint or bearing.

Such replacements can involve problems arising from stress redistribution when bridge beams/decks are propped to enable extensive repairs to be carried out to bearings. Movement joint assemblies and bearings are proprietary systems and advice from the suppliers is desirable.

6.19.4 Waterproof membranes

When investigations show that the watertightness of the deck membrane is defective or suspect, the only practical solution is to renew it. This is a major undertaking as it involves removal and replacement of the running surface and therefore serious disruption of traffic flow.

The detailing of the membrane at movement joints and at the longitudinal edges (kerbs) is of great importance to ensure watertightness. There are a number of membrane systems on the market, the majority of which hold Agrément Certificates. For highways, approval by the DoT is required.

The DoT approved list contains preformed sheeting and in situ compounds. The relevant DoT requirements are contained in relevant DoT publications to which reference should be made.

6.19.5 Strengthening with epoxy-bonded steel plates

In the early 1970s, the Transport and road Research Laboratory carried out a series of flexural tests to determine the effect of bonding external steel plates to the tension face of reinforced concrete beams. It was hoped that the tests would indicate the amount of composite action that would be developed between the steel plates and the concrete through the bond provided by the epoxy adhesive.

The use of this method of strengthening bridge decks and beams had the considerable advantage of minimum interference to traffic and where head-room was limited, the reduction caused by the bonding of the steel plate was insignificant.

The effect of long-term creep on the epoxy adhesive and the effect of weather and contamination by de-icing salts in surface run-off from bridges needed realistic assessment.

The use of epoxy-bonded steel plates to strengthen concrete structures, mainly bridges, had been successful in Japan, South Africa and Switzerland.

In addition to the work by TRRL, testing and development was carried out at the universities of Dundee, Sheffield and Warwick.

With this type of repair the critical factors are:

1. the preparation of the surface of the steel plate which is in contact with the epoxy adhesive;

2. the quality of the surface of the concrete beam soffit and the preparation of this surface to receive the epoxy adhesive;
3. the correct formulation of the epoxy adhesive for the site conditions under which it has to be used.

Also important is the preparation of the exposed surfaces of the steel plate and the application of a durable protective coating.

6.19.6 Post-tensioned bridges

The deterioration and repair of this type of bridge structure raises problems not associated with 'traditional' reinforced concrete. These 'special' features relate mainly to investigation of the condition of the prestressing cables and the anchorage blocks. Both the concrete and the prestressing operate at high stress and so defects can be more serious.

Each case presents a separate problem and this can only be resolved by a very experienced engineer and a very experienced contractor.

Repairs to the reinforced concrete units would follow standard good practice as previously described in this book.

As mentioned in section 6.11 ('Investigations'), it may be found that while the bridge was originally designed in accordance with the Code valid at that time, it no longer complies with present day Codes. The repair options would then have to be carefully considered and presented in detail to the client.

Fractures in wires of prestressing cables can be caused by corrosion arising from ingress into the ducts of water contaminated with de-icing salts, fatigue or stress corrosion.

6.19.7 Remedial work for ASR damage

As mentioned in section 6.18 the extent to which ASR affects the load-carrying and long-term durability of concrete structures is still open to some technical argument.

In an existing structure the only item of the three conditions mentioned which can be to some extent controlled/modified is the moisture content of the concrete. In theory, if the concrete could be dried out, the reaction would cease. In practice, all that can be done is to take measures to prevent or substantially reduce the continued ingress of moisture into the concrete. This would require a careful review and probable modification of the surface-water drainage system of the bridge and of the deck membrane, and the application of a durable coating. The coating should be specially selected to provide an effective barrier to the ingress of water and at the same time allow the passage of water vapour outwards from the concrete (i.e. it should allow the concrete to 'breathe').

PART 4 REPAIR OF CONCRETE SILOS AND BUNKERS

6.20 INTRODUCTION

Silos are usually circular on plan and can have a diameter/height ratio of 6 to 8, e.g. 6 m diameter and 35 m high, and are normally constructed in large groups.

Bunkers are often square on plan with a height about 1.5 times the base dimension.

The main document on the design and construction of silos is that produced by the ACI Committee 313 and is referenced in the 'Further reading' section at the end of this chapter. The term silo includes both deep and shallow bins; the latter is often referred to as bunkers. The publication emphasizes that the design must take into account both static and flow loading, and represents the latest thinking on both design and construction.

Both silos and bunkers are usually provided with a hopper-shaped bottom to facilitate emptying.

Reinforced concrete silos and bunkers have been in use for the storage of fine and coarse granular material for more than 70 years. During this period deterioration has usually occurred and repairs and sometimes strengthening are needed.

There are a number of causes for this deterioration, but the principal ones are:

1. Severe abrasion of the inside surface of the structures by the coarse-grained materials stored. Examples of coarse-grained materials are coal, clinker, limestone. In the case of clinker, this can contain chemical compounds potentially aggressive to Portland cement concrete, depending on the source material.
2. Corrosion of the reinforcement resulting from an aggressive environment porous concrete and/or inadequate cover to the reinforcement.
3. Some inadequacies in the original design which generally shows as vertical cracks.

6.21 Investigations

Investigations should follow the lines described in Chapter 4 and in Chapter 6, section 6.2, as deemed appropriate by the engineer. My experience is that many of these older structures make a first impression of being fit only for demolition, but a detailed investigation shows that such a drastic solution is not justified. (See Figures 6.7 and 6.8.)

Figure 6.7 Close-up view of crack in wall of grain silo. Courtesy, Gunac UK Ltd.

It is necessary to examine in detail the inside of the structure as well as the outside. It may be that the inside has suffered both abrasion and chemical attack (for the reasons given in the previous section) and this has resulted in rebar corrosion. The outside has probably suffered cracking and spalling also caused by rebar corrosion, and in some older structures by inadequate design. The stresses developed in these structures during filling and emptying are very complex and there are differences of opinion on certain aspects of design. In 1991 the American Concrete Institute produced recommendations for the design and construction of concrete silos for holding granular material; this publication has been referred to in the previous section.

Figure 6.8 Silo in Figure 6.7 under repair showing part of final treatment with three-coat reinforced elastomeric coating. Courtesy, Gunac UK Ltd.

6.22 EXAMPLES OF REPAIR

Methods of repair are similar to those described in Chapter 5 and in Part 1 of this chapter. Special care is needed in the selection of a curing method, as curing compounds should not be used if the silo is intended to hold edible materials. Similarly, linings to improve abrasion resistance and flow of materials in the silo require special selection.

A grain silo had a number of wide vertical cracks in the walls. The silo was approximately 100ft (33 m) high and 25 ft (8 m) diameter. This was one silo in a large group, the majority of which exhibited similar defects. They were all successfully renovated by cement/sand SRB mortar patch repairs and the provision of 75 mm thick reinforced gunite lining. The inside surface was prepared by grit blasting to remove all weak concrete and the reinforcing high-tensile mesh was securely pinned to the base concrete. Care was taken to ensure the embedding of the steel mesh in the gunite.

It is sometimes found that the original use of the silo has been changed and then the new use must be taken into account in selecting the repair method and materials. It is also prudent to check that the original design, either from original drawings or from a detailed site investigation, is adequate for the new use, as this may impose additional loading on the structure.

In an old silo used for limestone, it was found that in addition to general deterioration (abrasion of the concrete and corrosion of reinforcement), the slope of the hopper bottom was inadequate and considerable difficulties were experienced in emptying. It was therefore decided to increase the slope of the bottom to something over 50°. To do this in concrete would require a top shutter, but to place and properly compact concrete on such a slope under a top shutter is very difficult. The choice of method lay between using superplasticized concrete and shuttering, or shotcrete (pneumatically applied concrete). The selected method was heavily reinforced shotcrete, using a granite aggregate.

Figure 6.9 View of silos shown in Figures 6.7 and 6.8 after completion of repairs. Courtesy, Gunac UK Ltd.

Figure 6.9 shows repairs to a group of silos holding grain. On completion of the concrete repairs all the silos were finished with a French coating, IRTOP 1.4, applied in three coats incorporating a non-woven polyester reinforcing fabric in the intermediate coat.

6.23 REFERENCES

1. British Research Establishment (1991) Structural appraisal of existing buildings for change of use, *Digest*, **366**, Oct., 6.
2. Concrete Society (1976, addenum 1987), *Concrete Core Testing for Strength*, Technical Report No. 11, 44.
3. Dave Parker (1996) Sticking to the task, *New Civil Engineer*, 11 July, 22.
4. Department of Transport, Press Notice No. 260. See also paper by Dr Mark Raiss presented to the 74th annual meeting of the Transportation Research Board, Washington, DC, Jan. 1995. A shortened version of this paper appeared as S.M. Raiss (1995) Post-tensioned concrete bridges; the UK debate, *Concrete*, Mar./Apr., 23–6.
5. Hay and Virmani (1985) North American exposure in concrete bridge deterioration and maintenance Paper at 1-day symposium on concrete bridges. Programme: investigations, maintenance and repair. London, Sept.

6.24 FURTHER READING

General building structures

American Society for Testing Materials Designation E-119.88 (1993) *Standard Methods of Test of Building Construction and Materials*, Section 4, Vol. 4.07.

Bamforth, P. (1992) The development of cathodic protection systems, *Concrete*, Mar./Apr., 49–52.

British Standards Institution (1981) *Assessment of Concrete Strength in Existing Structures*, BS 6089.

British Standards Institution (1991) *Cathodic Protection*, BS 7361: Part 1: 1991, Code of Practice for land and marine applications.

Building Research Establishment (1991) Structural appraisal of existing buildings for change of use, *Digest*, **366**, Oct.

Building Research Establishment, (1989) *Static Load Testing of Building Structures*, Information Paper, IP/9/89.

Concrete Society (1988) *Concrete Core Strength Testing*, Digest No. 9, 10.

Concrete Society (1976, addendum 1987) *Concrete Core Testing for Strength*, Tech. Report 11, 44.

Darby, M. (1993) Realkalizing concrete, *Concrete Quarterly*, Spring, 21–2.
Levy, M. and Salvadori, M. (1992) *Why Buildings Fall Down – How Structures Fail*, W.W. Norton & Co.,l New York and London, p. 324.
McAnoy, R.P.L., Palmer, R.A. and Longford, P.E. (1987) Cathodic protection of reinforced concrete structures-experience in UK, Hong Kong, and Australia, *2nd International Conf. on Deterioration and Repair of Reinforced Concrete in Arabian Gulf*, Oct. Bahrain, pp. 121–38.
Moss, R.N. (1991) Load tests of floors and roofs, *The Structural Engineer*, **69** (19/1), Oct.
Parker, Dave (1996) Sticking to the task, *New Civil Engineer*, 11 July, 22.
Polder, R.B. Walker, R.J. and Page, C.L. (1995) Electrochemical desalination of cores from a reinforced concrete coastal structure, *Mag. of Concrete Research*, **47** (1973), Dec., 321–7.
Price, W. (1995) Potential strength, cores for concern, *Concrete*, Sept./Oct., 51–2.
Whittington, H.W., McCarter, T. and Forde, M.C. (1981) The conduction of electricity through concrete, *Mag. of Concrete Research*, **33** (110), Mar. 48–59.

Repairs to fire-damaged concrete buildings

American Society of Civil Engineers Structural Fire Protection (1992) *Manual of Practice*, 70.
Concrete Society (1990) *Assessment and Repair of Fire-damaged Concrete Structures*, TR. 033, 80.
Dorsch, D.F. (1993) Assessment and repair of fire-damaged concrete structures, *ASCE Proc. Materials Engineering and ASCE National Convention*, Dallas, Oct. 16–26.
Green, J.K. and Long, W.B. (1971) Gunite repairs to fire-damaged concrete structures, *Concrete*, April, 7.
Gumta, V.K., Roy, B.N., Kumar, V. (1995) Effect of fire on concrete structures and remedial measures: a review, *ICI Bulletin*, **50** Jan.–Mar., 25–37.
National Codes and Standards Council of the Concrete and Masonry Industries (1994) *Assessing the Condition and Repair Alternatives of Fire-exposed Concrete and Masonry Members*, PCA (USA), 14.

Repairs to concrete highway bridges

British Standards Institution Steel, *Concrete and Composite Bridges*, BS 5400: Parts 1: 1988; 2: 1978; 4: 1990; 7: 1978; 8: 1978; 9: 1983; 10: 1980.
British Standards Institution (1980) *Hot Rolled and Processed High Tensile Alloy Steel Bars for the Prestressing of Concrete*, BS 4486.
British Standards Institution (1980) *High Tensile Wire and Strand for the Prestressing of Concrete*, BS 5896.
British Standards Institution *(1991) Cathodic Protection*, BS 7361: Part 1: 1991, Code of Practice for land and marine applications.
Bamforth, P. (1992) Development of cathodic protection systems, *Concrete*. Mar./April, 49–52.
Broomfield, J. and Inam, J. (1992) Research on concrete and structures, Strategic

Highway Research Programme, *Concrete*, Nov./Dec., 19–24.
Cavalier, P.G. and Vassie, P.R. (1981) Investigation and repair of reinforcement corrosion in a bridge deck, *Proc. Inst. Civ. Engrs*, Part 1, 461–80.
Concrete Society (1989) *Concrete Bridges: Management, Maintenance and Renovation: Proc. of One-day Conference*, London, Feb., CS 102, 48.
Concrete Society (1995) *Alkali-silica Reaction: Minimising the Risk of Damage to Concrete*, TR 030, 48.
Concrete (1993) Grouting specifications, July/Aug., 23–8.
Concrete Society (1996) *Durable Bonded Post-tensioned Concrete Bridges*, Tech. Report 47, 72.
Dept. of Transport (1993) *Post-tensioned Concrete Bridges: Prioritisation of Special Inspections*, BA 54/93.
Dept. of Transport (1993) *Post-tensioned Concrete Bridges: Planning, Organisation and Methods for Carrying out Special Inspections.*
Dept. of Transport (1993) *Design Manual for Roads and Bridges*, Sect. 4., Part 3: Assessment BD 21/93.
Dept. of Transport (1994) *The Assessment of Concrete Structures Affected by Alkali–silica Reaction*, BA 52/94.
Dept. of Transport (1995) *The Assessment of Concrete Structures Affected by Steel Corrosion*, BA 51/95.
Dept. of Transport *The Assessment of Concrete Highway Bridges and Structures*; to be used in conjunction with BD 21.
Dereymaeker, J. (1993) *Structural Repair by Epoxy-bonded Plates*, Engineering Technical Press, 225–31.
Kendell, K. and Lewis, D.A. (1984) *Bridge Decks: Cathodic Protection*, Transport and Road Research Laboratory.
National Research Council (USA) (1981) *Concrete Sealers for the Protection of Bridge Structures*, National Cooperative Highway Research Program Report No. 244.
Raiss, M. (1995) Post-tensioned concrete bridges: the UK debate, *Concrete*, Mar./April, 23–6.
Raithby, K.D. (1980) *External Strengthening of Concrete Bridges with Bonded Steel Plates*, TRRL Supplementary Report 612, Transport and Road Research Laboratory, 28.
Robery, P. (1991) Specialist spraying key to silane success, *New Civil Engineer*, 6 June, 16.
Somerville, G. (1996) Grouted post-tensioned concrete bridges – the current position, *Concrete*, May/June, 22–4.
Vassie, P. (1984) Reinforcement corrosion and the durability of concrete bridges, *Proc. Inst. Civ. Engrs.*, Part 1, Aug., Paper 8798, 713–23.

Repairs to concrete silos, bins and hoppers

American Concrete Institute (1991) *Commentary on Standard Practice for Design and Construction of Concrete Silos and Stacking Tubes for Storing Granular Materials*, ACI 313R–91, 22.
American Concrete Institute (1991) *Standard Practice for Design and Construction of Concrete Silos and Stacking Tubes for Storing Granular Materials*, ACI 313–91, 27.
American Concrete Institute (1965) *Shotcreting*, Publication SP-14, 13 papers, 23.

Concrete Society (1979) *Specification for Sprayed Concrete*, ref. CS 021, 4.

Faber, J. and Mead, F. (1967) *Reinforced Concrete*, E & F.N. Spon, London, 330–52.

Rowe, R.E. (1958) *Investigation into Cracking in a Reinforced Concrete Cement Silo*, Technical Report 296, Cement & Concrete Assoc., April.

7

Coatings (barrier systems) for reinforced concrete

7.1 INTRODUCTION

In the UK, in the 1970s, it became clear that concrete was not proving as durable as originally expected, due mainly to the corrosion of the reinforcement. Since then a considerable amount of research work has been carried out in the UK, West Germany and the USA and numerous reports and papers have been published. The 'Further reading' section at the end of this chapter contains details of some of these publications. In the US, coatings for concrete are usually referred to as 'barrier systems'.

The main thrust of this research has been to develop equipment and methods for testing coatings for:

- water resistance;
- vapour transmission resistance;
- carbon dioxide (CO_2) transmission resistance;
- resistance to transmission of chlorides in solution;
- resistance to chemical attack;
- resistance to abrasion.

In the UK there is no specific Standard for testing coatings, but the recommendations in the appropriate Parts of BS 3900: Methods of Test for Paints, are used. The Parts generally included in specifications for coatings are:

Group E	Mechanical tests
Part E6	Cross-cut test
Part E10	Pull-off test for adhesion
Group F	(in 13 Parts); Durability tests on paint films
Group G	Environmental tests on paint films (including tests for resistance to corrosion and chemicals)

In Germany, there is a DIN 53 122: Determination of Water Vapour Transmission. Professor Klopfer of Dortmund is a recognized authority on carbon dioxide transmission through coatings, so much so that a proposed minimum resistance to the passage of CO_2 of 50 m of still air is often referred to as the 'Klopfer minimum'.

The effective useful life of a good coating properly applied to external concrete is likely to be about 15 years. Coatings should always be applied in a minimum of two coats, and the material should possess some degree of flexibility (i.e. it should be elastomeric). This is to reduce the risk of cracking due to temperature changes and inevitable movements which occur in all buildings. It is relevant that dark colour coatings absorb heat appreciably more than light colour ones. Reference can usefully be made to *BRE Digest* 227, 228 and 229: Estimation of Thermal and Moisture Movements and Stresses, Parts 1–3: 1979. Table 2, Digest 228 shows service temperature ranges in the UK: freestanding or fully exposed concrete structures, light colour: range –20°C to 45°C; dark colour –20°C to 60°C; thus giving a range of 65°C and 80°C respectively. In Table 1 in the same issue, the coefficient of thermal of concrete made with dense aggregate concrete (gravel and crushed rock) is in the range $10–14/10^{-6}/°C$. From this it will be seen that an rc beam 5.0 m long would have a potential thermal movement in the range of 3 mm to 4.5 mm with a light-coloured surface. For this reason, suppliers of coating materials should be asked to provide clear information on the elastomeric properties of their material. This is particularly important when the coating has to bridge cracks, some of which are not as 'dead' as they appear to be.

With any coating, it is prudent to inspect it at reasonable intervals, and to accept that patch repairs are likely to be necessary from time to time. The 'reasonable intervals' will largely depend on the type of coating and degree of exposure and accessibility; the range is likely to be 3–10 years.

Generally, coatings vary in thickness from about 0.20 mm to about 1.75 mm, depending on the type of material used, the number of coats and whether the coating is reinforced.

Economic considerations should not be allowed to influence unduly the selection of coatings. If a coating fails prematurely, the cost of renewal, possibly including the removal of the old coating, will cost considerably more than the original use of a more expensive, but more durable coating in the first instance.

Coatings used for aesthetic or decorative purposes only, are considered as paint systems and are not covered by this book; the subject is well covered by BS 6150: Code of Practice for the Painting of Buildings.

Many of the materials used in coating systems require special safety requirements to be taken in their use; the recommendations of the suppliers should always be followed.

7.2 COATING MATERIALS IN GENERAL USE

7.2.1 Material types

There is a wide range of coating materials on the market and these can be summarized as follows:

- bituminous – hot applied and cold applied;
- chlorinated rubber;
- polymers, water dispersed, solvent-free and with solvent:
    - acrylic – water-dispersed;
    - epoxies – solvent-free and with solvent;
    - polyurethanes;
    - silicones and silanes.

There are various combinations of these materials and selection should be based on a clear conception of the duties which the coating will have to perform, that is, the reason for specifying the use of a coating. This is dealt with in section 7.3.

7.2.2 Desirable characteristics

While these 'desirable' characteristics will naturally vary according to the use to which the coating is to be put, the main features for all coatings include the following:

1. The coating must be capable of forming a good bond to the substrate.
2. When applied externally, it must be resistant to weathering over many years, which means resistant to general deterioration, including chalking and loss of colour.
3. It must have high resistance to the passage of water.
4. It must possess low resistance to the passage of water vapour outwards from the substrate to avoid stress on the bond at the interface.
5. It should be resistant to the transmission of carbon dioxide; but see comments in section 7.5.
6. It should be elastomeric so as to avoid over-stressing due to thermal and other movements.

7.3 REASONS FOR USING COATINGS AS PART OF CONCRETE REPAIRS

The following are the principal reasons for using coatings on reinforced concrete:

1. as a waterproofing, carbon-dioxide-resisting, decorative finish;
2. in certain cases, the coating will be required to resist the ingress of chlorides in solution; examples are highway bridges, multi-storey car parks and structures near the sea;
3. the coating may be required to resist the passage of water under hydrostatic pressure, outwards as in water-retaining structures;
4. the coating may be required to assist in preventing the ingress of ground water when applied to the inside surface;
5. the coating may be required to protect the concrete from chemical attack;
6. the coating may be required to inhibit efflorescence on the surface of concrete, particularly when the concrete is pigmented.

Coatings under (5) above may also have to possess additional characteristics, such as resistance to elevated temperature and resistance to abrasion.

It should be noted that coatings used for many of the purposes referred to above, which are exposed to view, should also be aesthetically acceptable.

7.4 BASIC REQUIREMENTS FOR THE APPLICATION OF ALL COATINGS

The first requirement is to select the most suitable coating for the duty it has to perform.

For all coatings the basic requirement is to obtain the best bond between the substrate (concrete or cement-based mortar), and the coating.

Essentially, the precautions to be taken to achieve this bond are similar for all types of coatings and this consists in the careful preparation of the surface to which the coating will be applied. The actual details of the work required will depend largely on the condition of the surface. The surface must be clean and free of loose particles and for the majority of coating materials the surface should be dry. There are some specially formulated epoxy resins, and proprietary acrylic polymers dispersed in water, which can be applied to damp concrete and cement-based mortar.

There are occasions when a coating has to be applied to a rather weak concrete; this is likely to require a special primer or stabilizing liquid. The bond in such cases will not be as strong as when the coating is applied to high-strength concrete, as the bond at the interface cannot be stronger than the strength of the concrete surface to which the coating is applied.

The usual methods of surface preparation are:

1. cleaning the surface;
2. wire brushing;
3. light grit blasting;
4. high-velocity water jets; this can be the only practical method of removing old deteriorated elastomeric coatings; however, it leaves the surface layers wet which must be taken into account when selecting the coating;
5. the careful use of acid etching; this should only be used where other methods (as described above) are not practical. Special precautions are needed to protect both workmen, the public, and any exposed metal such as window frames etc. The acid-etched surface must be thoroughly washed down with clean water to remove all traces of the acid. This should be checked by the use of indicator paper which measures the pH. The pH should not be lower than 10.0.

It must be kept in mind that coatings are relatively thin, being in the range of 0.20 to 1.75 mm, and are expensive. The additional thickness required to cover high points fully on a rough surface created by the over-enthusiastic preparation achieves only unnecessary expense.

7.5 COATINGS FOR USE AFTER GENERAL REPAIRS TO CONCRETE

The use of these coatings does introduce a maintenance commitment: the useful life of such coatings should not be expected to exceed about 15 years, depending on exposure conditions, the type of coating and its thickness.

It is inevitable that fine shrinkage cracks develop around the perimeter of repaired areas; also the new concrete or mortar used for the repair will not match the existing concrete. A high-quality coating, properly applied, will deal with both these problems and will ensure a significant increase in the useful life of the repaired areas. The whole of the repaired areas and the adjacent areas which have not been repaired should be sealed with the selected coating. This will reduce the adverse effects of low cover and permeable concrete for all the units dealt with. (See Figures 7.1 and 7.2).

When considering a coating for use after repairs to a concrete building, the following criteria should be met:

1. A good 'track record' of previous performance; the suppliers of apparently suitable materials should provide details of work carried out with dates. This will enable colour retention and general weather resistance to be checked.

Figure 7.1 View of block of flats with defects in the concrete, and in the cement/sand rendering and in the mosaic finish. Courtesy, Gunac UK Ltd.

2. It is now standard practice for suppliers to be asked to provide test information on the resistance to diffusion through the coating of carbon dioxide (CO_2). I have serious reservations about the value of this test for the following reasons:

   (a) The weathering of the coating has a significant effect on the resistance to diffusion of carbon dioxide. So much so that tests by BRE showed a dramatic reduction in the carbon dioxide resistance after three years' natural weathering and 2000 hours' artificial weathering.

Figure 7.2 View of flats in Figure 7.1 after repair and finished with reinforced elastomeric coating. Courtesy, Gunac UK Ltd.

(b) Some coatings which possessed a very high resistance to CO_2 transmission when new (figures of up to five times the Klopfer minimum of 50 m of still air were recorded) showed a loss of about 80% after three years' natural weathering.

The above results throw into serious doubt the value of the present emphasis on testing coatings for CO_2 transmission.

The emphasis on the need to reduce the diffusion of CO_2 into the concrete is to reduce or eliminate the danger of concrete carbonating down to the rebars and thus leading to loss of passivation of the steel rebars. For corrosion to take place, both oxygen and moisture must be present.

In view of the problems outlined above, I have concluded that performance criteria for coatings applied externally to reinforced concrete after repairs should be based on the ability to reduce water penetration, the possession of a high degree of vapour transmission, very good bond with the substrate and good elastomeric properties to prevent over-stress due to building movement.

7.6 COATING OVER CRACKS

The presence of cracks in members can present a serious problem and this has been discussed in Chapter 4, section 4.6 and Chapter 5, sections 5.3.7 and 5.3.8.

In considering the type of coating required, the first problem is whether the cracks are 'live' or static. This can be difficult to decide as movement across the cracks can be small. As the application of the coating is generally the final stage in the repair programme, it is assumed that a decision about the nature of the cracks will have been taken when the general concrete repairs programme and specification was prepared.

My experience is that it is desirable for **all** coatings to possess some degree of flexibility and the width of crack which the coating can bridge safely should be stated by the supplier.

Generally, the wider the crack the thicker the coating: a high-quality coating applied in three coats should be able to bridge cracks up to 1.00 mm wide. Cracks wider than 1.00 mm wide would require three coats with a selected fabric reinforcement in the second coat. This is emphasized in a range of French coatings (IRTOP).

Transverse cracks in cantilever balcony slabs can be particularly difficult to deal with unless the top surface of the concrete is accessible so that the cracks can be repaired there, rather than from the underside. If these cracks can be correctly repaired, then my experience is that they can be considered as static and the coating selected accordingly. If they cannot be correctly repaired then water will penetrate through from the top surface and accumulate behind the coating, forming large blisters. See Chapter 6, Part 4, Figures 6.7–6.9.

7.7 COATINGS TO RESIST THE INGRESS OF CHLORIDE IONS

Papers on chloride attack on steel rebars invariably refer to the 'diffusion of chloride ions' into concrete. Chlorides in solution are in the form of ions. A coating which will successfully resist the passage of water will equally well resist the passage of solutions of chlorides, unless the coating material itself is vulnerable to attack by the chlorides. For example, a coating may be very resistant to a solution of calcium or sodium chloride, but may be vulnerable to ammonium chloride. It can be very difficult to assess accurately the depth of penetration of water into concrete, but the presence and concentration of chlorides in concrete can be detected by standard test methods.

Such coatings are likely to be needed for highway bridges due to the use of deicing salts. In the case of multi-storey car parks, cars bring in an appreciable amount of water containing high concentrates of de-icing salts. This is a major problem in the US and Canada and numerous reports have been written on the subject. It has occurred in the UK but to a lesser extent.

In the case of highway structures, it may in some cases be desirable for the coating to possess a high resistance to abrasion due to surface water containing deicing salts and grit being thrown against the concrete.

The Department of Transport (DoT) have a specification for silanes for use on concrete highway bridges. Silanes can be described as hybrid molecules composed of organic and inorganic parts. The size of silane molecules is in the range of 5–10 A, about the size of water molecules. This enables the compound to penetrate into the surface layers of the concrete and not just form a well-bonded surface layer. There are various types of silanes and it appears that the alkylsilanes are best suited to resist the penetration of solutions of chlorides. I do not consider silanes are suitable when the surface of the concrete has to be protected against abrasion and/or chemical attack.

The parts of concrete marine structures above low water level up to several metres above high water level (the splash zone) are vulnerable to abrasion and to the ingress of sea water which contains chlorides. In Atlantic waters this is about 18 000 ppm chlorides and in the Red Sea and Arabian Gulf about 22 000 ppm. Therefore it is doubtful if any of the coating materials considered in this chapter would be suitable for long-term performance under marine conditions. As mentioned below in Chapter 9, Part 3, I recommend the application of gunite, not less than 60 mm thick, reinforced with a galvanized steel mesh.

7.8 COATINGS (BARRIER SYSTEMS) TO PROTECT CONCRETE AGAINST CHEMICAL ATTACK

It is obvious that the coating selected must itself be immune to attack from the chemicals against which it has to protect the concrete. In many cases, the coating must also be able to withstand elevated temperatures. This latter requirement can greatly restrict the choice of coating and in extreme cases may require the use of chemically resistant ceramic tiles bedded and jointed in chemically resistant mortar.

Information on chemical attack on concrete is given in Chapter 3, section 3.5, and basic characteristics of epoxy, polyester and polyurethane resins are described in Chapter 2, section 2.14.

The most aggressive chemicals are probably mineral acids. A substantial barrier of inert material is therefore recommended. Generally, epoxies, polyesters and polyurethanes are resistant to a wide range of acids. Some general information on these three basic resins is given in Chapter 2, section 2.14 to which the reader is referred. It is important to remember that epoxies are in many types; polyesters are also in many types, but only two are used for protective barrier systems; polyurethane resins for barrier systems are essentially based on two reacting polymers, one of which is the curing agent. The two principal resins used are epoxies and polyurethanes.

For mild conditions they would be applied up to 1.00 mm thick as two-pack materials containing fillers; application by spray or brush, in two coats plus an appropriate primer. If brush-applied, the second coat should be at right angles to the first coat to help to eliminate pinholes.

Good bond to the substrate is of great importance.

A relevant publication is the American Concrete Institute's Manual of Concrete Practice, ACI 515, IR-79 (revised 1985) containing *A Guide to the Use of Waterproofing, Dampproofing, Protective and Decorative Barrier Systems for Concrete*. This document contains a comprehensive list of chemicals which attack concrete (Table 2.5.2.).

7.9 COATINGS TO INHIBIT THE FORMATION OF EFFLORESCENCE

Efflorescence is the deposit of salts, whitish-grey in colour, on the surface of concrete. These salts emerge in solution from within the concrete and crystallize on evaporation. While all concrete is liable to efflorescence, the use of coatings to inhibit it is normally restricted to precast concrete, especially roof tiles. This type of coating is generally not intended to have a life exceeding about three years and is often acrylics in a dilute dispersion in water.

It is usually applied as soon after manufacture of the units as possible and helps long-term curing. It is often colourless, although by the time it wears off, some yellowing is likely to have occurred.

7.10 REFERENCE

1. Building Research Establishment (1993) *The Effect of Weathering on the Carbon Dioxide Resistance of Coatings*, BRE Occasional Paper, April, 27.

7.11 FURTHER READING

American Concrete Institute (1979, revised 1985) *A Guide to the Use of Waterproofing, Dampproofing, Protective and Decorative Barrier Systems for Concrete*, ACI, 515, IR-79, 43.

Building Research Establishment (1993) *The Effect of Weathering on the Carbon Dioxide Resistance of Coatings*, Occasional Paper, April, 27.

Concrete Society (1987) *Permeability Testing of Site Concrete*, Technical Report 31, 95.

Construction Research and Information Assoc. (1987) *Protection of Reinforced Concrete by Surface Treatments*, Tech. Note 130, 72.

Robery, P. (1991) Specialist spraying: key to silane success, *New Civil Engineer*, 6 June, 16.

8

Repairs to concrete floors and roofs

INTRODUCTION TO PARTS 1 AND 2

The repairs described in this chapter deal with the repair of concrete floor slabs, including cement-based bonded toppings where these are used as the wearing surface. Such floors are normally used for commercial and industrial purposes. Repairs and replacement of floor finishes such as screeds, tiles, vinyl sheeting etc. are not included.

Repairs to concrete roofs include remedying defects in the waterproof membrane. It is assumed that the roofs are 'flat'.

PART 1 REPAIRS TO CONCRETE FLOORS

8.1 INVESTIGATIONS

There are likely to be two principal reasons for an engineer's being requested to inspect and report on a concrete floor:

1. because the user is dissatisfied with the floor surface, for a variety of reasons;
2. because an opinion is needed on the likely durability of the floor due to change in use.

The complaint regarding floor surface can vary from dissatisfaction with small colour differences in a granolithic topping to serious cracking and disintegration of the floor surface and moisture penetration from ground water.

The engineer would be wise to deal with unjustified complaints clearly, politely and firmly.

When there are definite defects, it is necessary to establish whether, if the floor is at ground level, it was designed as 'uniformly supported on the ground' or as a suspended slab. This will affect the diagnosis of cracks and how they should be dealt with.

Some ground floor slabs are, for good reasons, designed as suspended slabs.

Floors above ground level are designed as suspended slabs supported on beams or load-bearing walls.

Information on the age of the floor and 'history of use' will be required for diagnosis and repair recommendations.

It is my experience that significant defects are likely to fall into the following categories:

(a) wear of the surface resulting in defective areas;
(b) crazing or 'map cracking';
(c) a smooth, hard surface which has become slippery and thus potentially dangerous to those using the floor;
(d) cracks – other than 'crazing'.
(e) defective joints, usually in the form of breakdown of the arrisses;
(f) defective areas caused by chemical attack arising from spillage of aggressive chemicals on the floor;
(g) in the case of wet trades on suspended slabs, seepage through the floor; this is seldom a problem in slabs supported on the ground, unless the seepage is excessive;
(h) in the case of ground-supported concrete floors, signs of dampness (moisture rising from the ground) should be looked into as this would indicate absence of, or damage to, the waterproof membrane;
(i) debonding/delamination of 'bonded' cement-based toppings.

In certain cases it may be required to investigate the thermal insulation of a ground-supported floor slab to check compliance with the Building Regulations 1991, Approved Document L: Conservation of Fuel and Power.

8.2 DIAGNOSIS AND RECOMMENDATIONS FOR REPAIR

The visual examination would form the basis for a detailed investigation which should enable the defects to be diagnosed and recommendations prepared for remedial work. There are so many factors involved in the deterioration of concrete floors that only an outline can be given here and this will deal with each of the nine types of defects more commonly found; listed above as (a) to (i).

8.2.1 Surface wear/abrasion

This defect is very seldom found uniformly distributed over the whole floor area, but is confined to a number of specific areas.

Abrasion may be due to the floor's being used for a purpose for which it was not originally intended. For example: originally intended for light commercial use (printing), actual use heavy mechanical workshop.

If this does not apply, then the concrete may be sub-standard with respect to abrasion resistance. The use of the rebound hammer can provide a very useful indication of the surface hardness of the concrete from which the likely compressive strength can be deduced. As discussed in Chapter 4, section 4.4.10.4, this should not be used as a pass or fail test for compressive strength, but is much more conclusive for surface hardness. When the results are unsatisfactory or inconclusive, to reach a final conclusion may require cores to be taken and tested for compressive strength, aggregate type and approximate grading, cement content and an opinion given on degree of compaction and probable standard of curing.

Information on core cutting and testing is given in Chapter 6, section 6.2.5.

It is particularly important to obtain information about the concrete in the undamaged areas as well as the damaged areas as this will affect the type of repair required.

8.2.1.1 Patch repairs

If it is concluded that the floor as a whole has reasonable wear resistance, then patch repairs to the damaged areas would be adequate. The materials used for the repair will depend largely on the depth of defective concrete which has to be removed. Where the depth is between 40 and 60 mm, then a cement-modified mortar can be used, the work being carried out as described below:

1. All weak and defective concrete should be cut away and resulting grit and dust removed. The patches should be cut in approximate squares or rectangles, as shown in Figure 8.1.
2. The surface of the concrete around the perimeter of the cut-out areas should be cleaned and wire brushed for a distance of 50 mm.
3. The surface of the cut-out concrete should be well wetted, preferably overnight, and then not more than 20 minutes before the new concrete or mortar is laid. A coat of Portland cement/SBR grout should be well brushed into the prepared surface of the cut-out concrete.
4. The new concrete/mortar must be well compacted and finished to give the required texture.

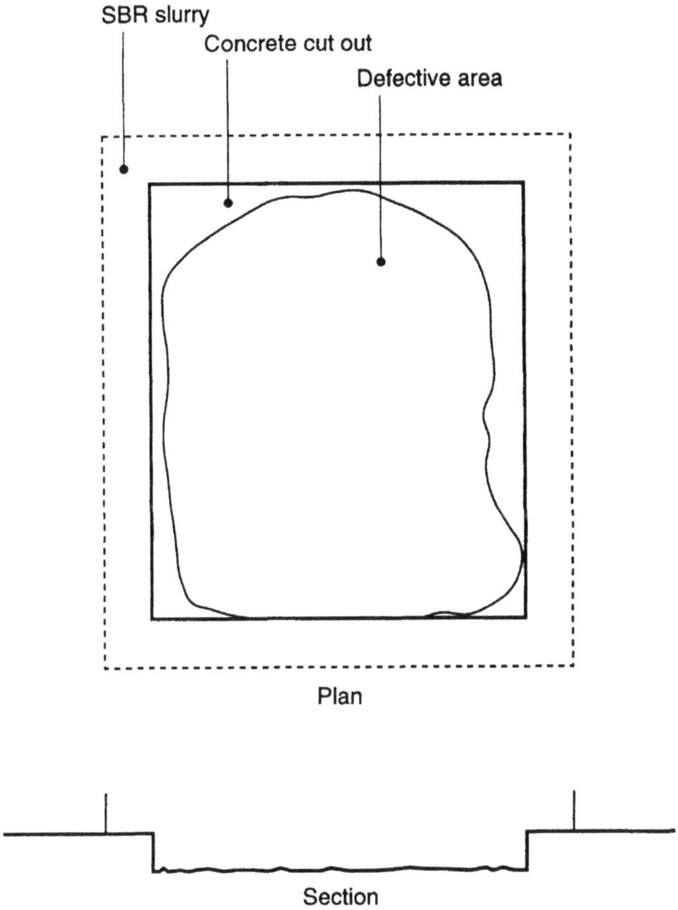

Figure 8.1 Sketch showing patch repair to concrete floor.

5. After the finishing operations are complete, the surface is covered with polyethylene sheets held down around the perimeter and kept in position for four days, or cured by the application of a spray applied curing membrane.
6. The mix proportions for the materials referred to above are given below:

Grout:	50 kg OPC to about 25 litres of SBR emulsion.
Concrete:	1 part OPC or RHPC to 2½ parts clean concreting sand, to 2½ parts 10 mm coarse aggregate, with 10

litres of SBR emulsion to 50 kg cement. The amount of added water should be just sufficient to provide adequate workability for compaction. The proportions given are by volume.

Mortar: 1 part cement, as above.
3 parts clean concreting sand.
10 litres SBR emulsion to 50 kg cement.
If prebagged sand is used the proportions by mass would be 1 part cement to 3½ parts sand.

If OPC is used, traffic should be kept off the repaired areas for seven days; if RHPC is used this can be reduced to two days. In cold weather, these 'keep-off' periods should be increased by 50%.

For emergency repairs, say at a weekend, high alumina cement (HAC) can be used in the same proportions as given above; also the SBR can be used to improve workability with a low w/c ratio which must not exceed 0.40. Wet curing is preferred if HAC is used.

Reference should be made to Chapter 2 for more information on HAC. If HAC is used for the mortar/concrete it must also be used for the slurry coat; it must not be mixed with Portland cement.

After the completion of the repairs it is advisable to apply a silicofluoride hardener to the surface of the repaired areas and the 50 mm wide prepared strip around the perimeter.

For patches in excess of about 60 mm deep, it is better to use concrete with 10 mm aggregate than a mortar. The mix proportions for the concrete would be 1:1:2.5 with 10 litres of SBR to 50 kg cement.

For patches less than 40 mm deep, it is usually better to use a proprietary prepacked mortar based on a reactive resin, of which there are a number on the market. The work should be carried out as described by the suppliers.

8.2.1.2 New topping or over-slab

If the investigation shows that the whole or a large part is unsuitable for the use to which it is or will be put, then there are really three alternatives:

1. Remove the whole floor and relay it to an appropriate specification. This is only practical for slabs on the ground. **Or**:
2. Overlay with a proprietary bituminous-based topping. Such toppings are composed essentially of bituminous emulsions, Portland cement, sand and chippings. They can be considered as semi-rigid and provide a virtually jointless, dustless, resilient and quiet surface, with a probable 'life' of about 20 years. Fine cracks are normally self-

sealing and wide cracks and damaged areas can be readily repaired. They are supplied and laid by such firms as Amey Flooring Limited and Colas Building Products. They are susceptible to attack by petroleum oils and various solvents. **Or:**

3. Lay a new concrete slab on top of the existing one with a slip membrane at the interface of the two slabs. This is known as over-slabbing. Such over-slabs should not be less than 100 mm thick and generally 125 mm is considered advisable.

The specification for the new slab and for the over-slab will depend on the use to which the floor will be put. For a slab where high abrasion resistance is required, the following is recommended:

- minimum Portland cement content: 350 kg/m^3;
- maximum additions (ggbs to BS 6699): 50 kg/m^3;
- maximum free water/cement ratio: 0.40;
- minimum characteristic compressive strength measured on 100 mm cubes: 50 N/mm^2 (BS 1881, Parts 108 and 116);
- aggregates to BS 882; flakiness index not to exceed 35 and aggregate impact value should not exceed 25%
- minimum characteristic strength of cores (should these be required): 30 N/mm^2 (BS 1881: Part 120);
- the slump would depend on method of compaction; with a w/c ratio of 0.40, a plasticizer would be required;
- thorough compaction is essential, as is proper curing with a resin-based spray-applied membrane.

It should be noted that a thin-bonded cement-based topping is not recommended where the base concrete is of poor quality. Also, the laying of new toppings can create problems with levels of adjacent floors, and the additional dead load of the new topping must be taken into account when the base slab is suspended.

8.2.2 Dampness in ground-supported floor slabs

This condition needs attention for two major reasons:

1. It constitutes a health hazard in occupied buildings and is contrary to the requirement of the Building Regulations 1991 and Section 3 of Approved Document C, 1992 edition.
2. Moisture rising from the ground/fill beneath the slab may bring with it sulphates in solution which over a period of years can cause serious damage to the concrete slab.

Thus, when dampness in a ground-floor slab is discovered, the possibility of sulphate attack should be investigated by taking cores through the slab into the ground/fill. Chemical analysis would enable a diagnosis to be made.

If sulphate attack has started then renewal of the slab is the only solution. It may also be necessary to remove and replace the fill beneath the slab.

The new floor system would then include a damp-proof membrane and possibly thermal insulation.

It can be difficult to provide a new damp-proof membrane in an existing floor without renewing the floor because the membrane has to be laid on the existing concrete slab and then the floor finishes, if any, on top of the membrane. If there are no floor finishes, as in a workshop, then the membrane has to possess high resistance to wear and abrasion and good bond to the concrete slab so as to resist vapour pressure from trapped moisture in the floor slab.

The insertion of a membrane into an existing floor system can create problems of loss of bond, either at the base concrete-membrane interface or at the membrane-surface finish interface, or both, depending mainly on the type of membrane used. The membranes in general use are:

Sheet membranes:

- simple polyethylene sheeting (1000 gauge minimum);
- high-density polyethylene sheet with rubber-bitumen adhesive on one side, e.g. bituthene;

In situ compounds:

- 'Synthomer' which is a liquid vapour membrane based on a water-based styrene butadiene latex which dries to a tough water- and vapour-resistant film. The wet film thickness should be at least 1.1 mm, applied in a minimum of two coats. It will bond well to a sound concrete base slab and the dry film will also bond well to an SBR modified cement/sand mortar topping.
- 'Ventrot' is a hot applied bituminous dpm laid by brush or spray to a nominal thickness of 3 mm on a special primer. Due to the hot bitumen base, this proprietary system requires special precautions to be taken.

8.2.3 Thermal insulation

If major replacement work is required as described above, it would be prudent to assess the thermal insulation of the floor in terms of 'U' value. This applies to floors on the ground. Reference should be made to the Building Regulations 1991, and Approved Document L – Conservation of Fuel and Power, Table 1 requires a maximum U (W/m^2K) of 0.45 for ground floors of dwellings and all other buildings.

If the existing floor slab is being removed and replaced with a new one then the thermal insulation can be laid below the slab with a new damp-proof membrane on top of the insulation which must be unaffected by moisture, e.g. 'Floormate' made by Dow Construction Products.

If this is an over-slabbing job, then the insulation would be located on a new damp-proof membrane laid on the existing concrete slab.

The above should be designed to meet the Building Regulations 1991 for both thermal insulation and damp-proofing.

8.2.4 Crazing or 'map cracking'

The effect of this type of cracking on the ability of the floor surface to resist wear is often misunderstood. Crazing can and does occur in hard abrasion-resistant concrete floor surfaces and does not, in most cases, result in premature wear. However, crazing may not be acceptable in a floor of a food processing factory and similar establishment where the risk of harbouring micro-organisms cannot be tolerated.

The cracks are very fine, usually in the range of 0.1 to 0.3 mm in width and up to about 2 mm deep. The main cause of crazing is over-trowelling; other causes include the use of an aggregate containing an excessive amount of very fine material, and/or the use of too much water in the mix.

A suitable treatment is to clean the surface of the concrete and apply a silico-floride hardener, such as 'Lithurin'.

8.2.5 Slippery floor surface

A high-quality, abrasion-resistant concrete floor sometimes become slippery and this can cause consternation to the users of the floor and risk of injury to the workers.

Another major cause of slippery floors is the presence of water on the floor resulting from the use to which is put, for example, a 'wet' trade such as food processing and dairies, and the parking areas in multi-storey car parks.

The resistance to slip is the friction at the interface of the two materials, the concrete and the soles of the footwear used by persons walking on the floor.

Changes in the coefficient of friction are caused by:

1. polishing of the surface under the action of traffic;
2. accumulation on the floor surface of material arising from the work being carried out, e.g. fine powder, wax, grease, oil etc.;

3. changes in the type of footwear used by the workers;
4. some combination of (1), (2) and (3).

There is an obvious conflict between the need for a slip-resistant surface and the need for a surface which can be cleaned easily; this is particularly so in food processing establishments.

Measures to remedy conditions under (2) and (3) are obvious. To remedy a polished concrete surface is not easy and practical measures are confined to mild acid etching and very carefully executed light grit blasting and the application of a fine abrasive grit.

When the cause of the slipperyness is the accumulation of water on the floor surface, major remedial work is likely to be required. This work is described later in this chapter in section 8.3.1.1. For more information on slippery floors, reference should be made to the 'Further reading' section at the end of this chapter.

8.2.6 Cracks

The presence of cracks in a concrete floor generally causes serious concern to the building owner or occupier, but this anxiety may be exaggerated. It is important to establish their long-term effect on durability, i.e. the ability of the floor to perform the function for which it was originally intended or for any proposed change in use.

Clearly defined cracks (other than those defined as 'crazing') usually arise from drying shrinkage and/or thermal contraction. Cracks can also arise from joints not being provided where needed, such as stress relief joints and isolation joints. Very occasionally they are found to be due to inadequate design which can result in unacceptable deflexion in suspended slabs.

In the case of serious cracking, information on the original design of the floor is most desirable; when this is not available detailed, expensive and time-consuming investigations are likely to be required. Reference can be made to Chapter 6, section 6.2 for more information on this subject.

Apart from their unsightly appearance, the main reasons for the repair of cracks are:

1. To prevent fretting and ravelling along the edges of the crack. Depending on the type of traffic using the floor, pedestrian, hard-wheeled trolleys, heavy fork-lift trucks. Damage to the joint edges is unlikely to occur with cracks less than 1 mm wide.
2. To seal permanently the cracks prior to the laying of a thin-bonded topping. This will not be effective if the cracks are 'live', as movement across the crack will crack the topping and may result in a limited amount of debonding alongside the cracks.

3. To prevent the penetration of water or other liquids through a suspended floor, particularly when the space below the floor is used.

If it is decided that the cracks are 'live' (in which case they would most likely be more than 1 mm wide), they should be cut out and sealed with a flexible sealant. Cracks caused by the inadequate provision of joints should be considered as 'live' and dealt with accordingly.

Cracks not exceeding 1.5 mm wide need not be cut out; it is recommended that the edges of the crack be tapped with a chisel and light hammer to detect any weak places, then the crack cleaned out with compressed air and filled with a semi-flexible sealant, such as an epoxy polysulphide.

Wide cracks require more detailed attention. The decision on repair would depend on the type of crack, its width and depth and whether rebars crossing or parallel to the crack are corroded. If corrosion has occurred, then a check must be made for presence of chlorides and if chlorides are the main factor in the corrosion consideration should be given to either cathodic protection or chloride extraction; see Chapter 6 for information on these two systems.

Crack injection may provide the answer (see Chapter 6). If the crack is assessed as being 'live', then a flexible sealant would have to be used after the crack had been prepared and cleaned.

If further movement across the crack is not anticipated, then cutting out and repair with mortar as described above in section 8.2.1.1 should be adequate.

8.2.7 Defective joints

In slabs on the ground there are four main types of joint:

1. full movement joints (also known as expansion joints);
2. contraction joints and stress-relief joints;
3. warping joints;
4. isolation joints.

Not every ground floor slab has all four types, as the type and number of joints depends on the design of the floor. Suspended slabs may have only day-work joints which are intended to be monolithic. In suspended slabs of large area, there may well be structural (full) movement joints as part of the design of the building.

Joints generally are a potential source of trouble, particularly full movement joints.

Damaged arrisses can be repaired by cutting away defective concrete and making good to the sides and edges with an epoxy mortar and then resealing the joint groove with a suitable sealant. Information on sealants is given in Chapter 2, section 2.13. Unfortunately, the possibility of the arrisses again breaking down still remains but this can be resolved by fixing across the joint a stainless-steel cover plate. It is advisable to fix the cover plate on one side of the joint only as this will allow movement to take place without interference from the cover plate.

The cover plate should extend 125 mm beyond the edge of the joint on the fixing side and 75 mm on the other (the 'free') side.

For expansion (full movement) joints where appearance is important, special proprietary joint assemblies are available and give satisfactory service, but they are expensive compared with the simpler solution outlined above.

A serious defect in full movement joints in ground-supported slabs can occur when dowel bars (provided for load transfer) are displaced when the slab is cast. This results in the locking of the joint causing cracking and damage to the concrete each side of the joint.

The solution is to remove both bays and replace taking special care with the location and fixing of the dowel bars.

Contraction, stress-relief, warping and isolation joints are quite narrow, the width being in the range of 5–10 mm.

The principal defect in these joints is damage to the arrisses. This can be repaired as suggested above, or the arris can also be protected by the installation of a steel angle, but this is difficult to fix securely in an existing joint.

8.2.8 Defective areas caused by chemical attack

Chemical attack on concrete floors usually arises from the spillage of aggressive chemicals. There is a discussion in Chapter 3, section 3.5 on chemical attack on concrete and reference is made to the compounds in general use which are aggressive to concrete.

Mineral acids, sulphuric, hydrochloric and nitric are particularly aggressive to concrete as the attack starts immediately the acid makes contact with the concrete and it continues until all the acid is neutralized by the highly alkaline cement paste. For this reason it is possible for great damage to be done in a short space of time.

There are three solutions:

1. to renew the damaged concrete as required;
2. to repair the damaged areas and protect the repaired area(s) with a coating or thin bonded topping which is inert to the acid used;
3. to take steps to prevent further spillage and repair the damaged areas.

The important factor here is to ensure that **all** defective concrete is removed before new concrete or mortar is laid. The details of the method of repair have been described in paragraph 8.2.1.1 (patch repairs), for general repairs. In the present case it is essential that all concrete attacked by the acid should be completely removed. When all visibly affected concrete has been removed, it is advisable for the surface of the existing concrete to be checked for alkalinity by means of indicator papers which should show a pH not lower than 10.0.

Information on protective coatings (barrier systems) has been given in Chapter 7.

8.2.9 Debonding/loss of adhesion of toppings

The toppings considered here are known as 'thin-bonded' toppings such as granolithic and high-strength toppings using selected aggregates, and laid for industrial and commercial use.

With such toppings considerable reliance is placed on achieving good bond with the base concrete.The loss of bond is shown as hollow-sounding areas when the topping is struck with a light hammer or rod. If such an area is subjected to heavy impact there is a danger that the topping will crack, often radially from the point of impact. These areas of poor adhesion often occur at the corners of bays and alongside shrinkage cracks.

Loss of bond/adhesion does not mean that the topping will necessarily fail, but it is a danger signal that serious defects may occur. Assessment of the situation is difficult and points to consider include:

1. use to which the floor is, or will be put;
2. thickness of the topping and its general quality;
3. the location of the debonded areas in relation to the traffic on the floor; areas adjacent to a wall are less vulnerable than in a doorway or on a gangway or main traffic route;
4. the degree of debonding and the percentage area of the floor affected.

When it is considered that the loss of adhesion may be serious and the floor area is large, then the use of an impulse radar survey is worth considering. A preliminary investigation, based probably on a 2.0 m grid would normally be required and this can cover about 3000 m^2 in a day. For detailed work, the grid would be reduced to 0.50 m centres.

8.2.10 Defects in the ground support to industrial floors

These floors are usually heavily loaded, due to moving loads and 'point' loads. They are normally designed on the assumption that the reinforced

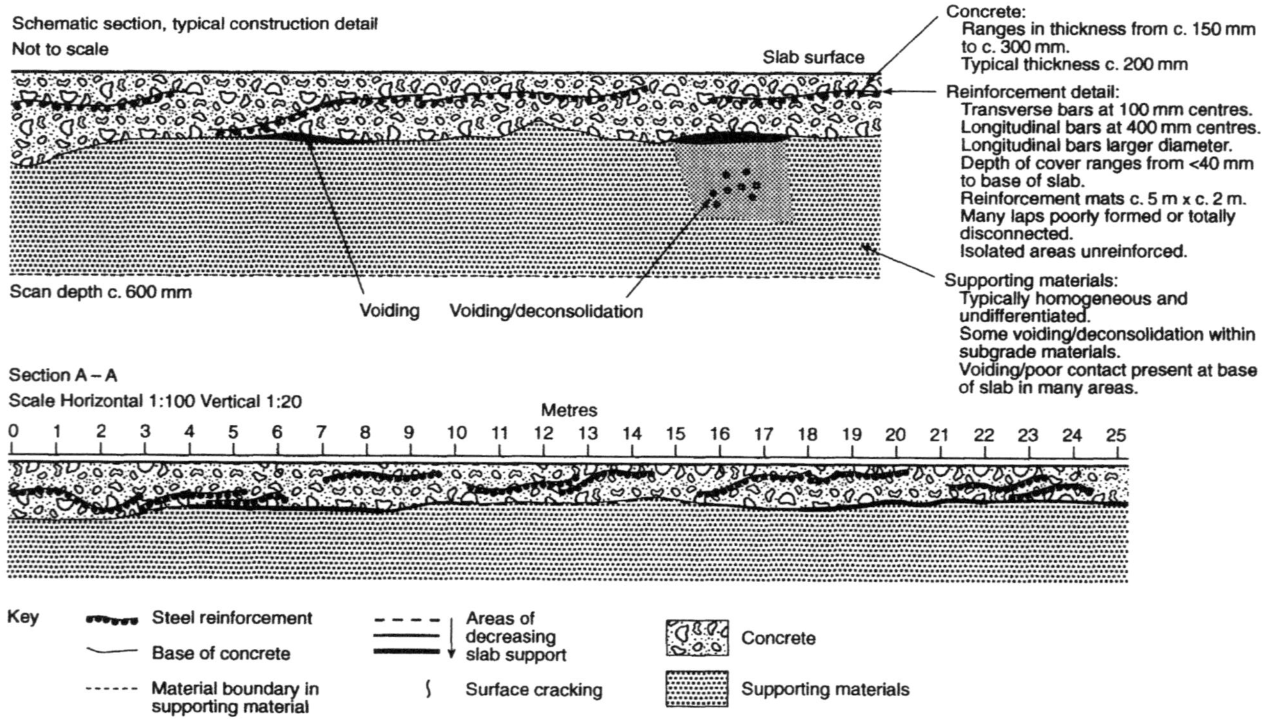
Schematic section, typical construction detail
Not to scale
Slab surface
Concrete:
Ranges in thickness from c. 150 mm to c. 300 mm.
Typical thickness c. 200 mm
Reinforcement detail:
Transverse bars at 100 mm centres.
Longitudinal bars at 400 mm centres.
Longitudinal bars larger diameter.
Depth of cover ranges from <40 mm to base of slab.
Reinforcement mats c. 5 m x c. 2 m.
Many laps poorly formed or totally disconnected.
Isolated areas unreinforced.
Supporting materials:
Typically homogeneous and undifferentiated.
Some voiding/deconsolidation within subgrade materials.
Voiding/poor contact present at base of slab in many areas.
Scan depth c. 600 mm
Voiding
Voiding/deconsolidation
Section A – A
Scale Horizontal 1:100 Vertical 1:20
Metres
0 1 2 3 4 5 6 7 8 9 10 11 12 13 14 15 16 17 18 19 20 21 22 23 24 25
Key
Steel reinforcement
Base of concrete
Material boundary in supporting material
Areas of decreasing slab support
Surface cracking
Concrete
Supporting materials

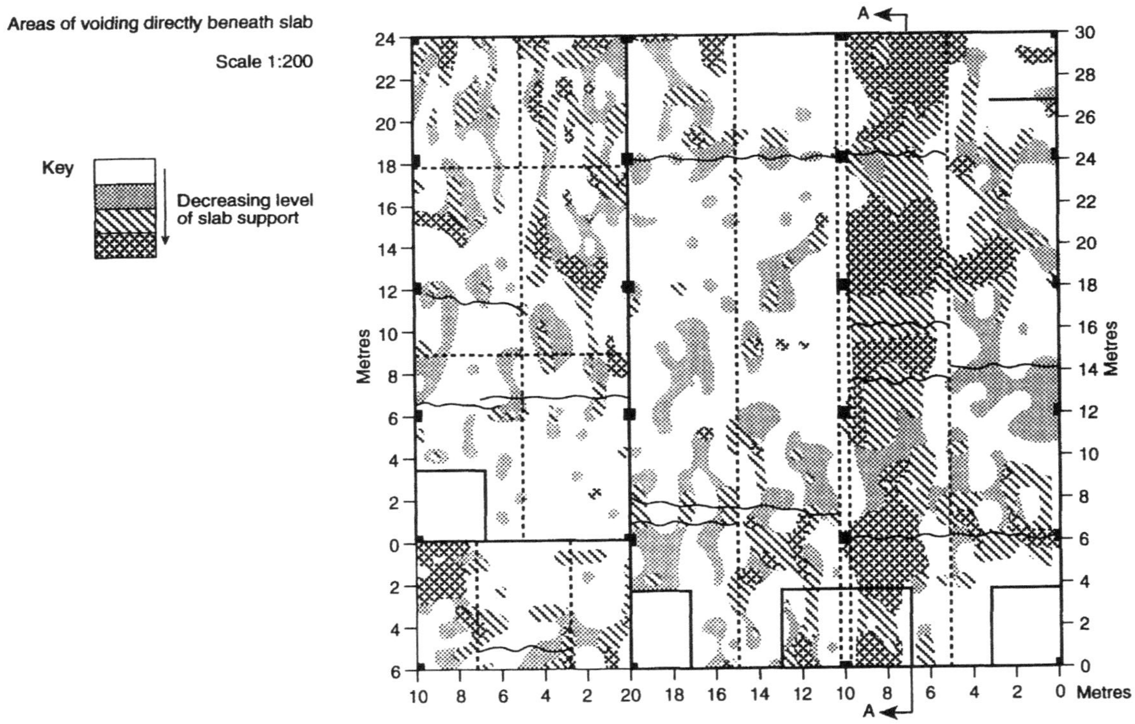

Figure 8.2 Result of investigation by impulse radar, of support to industrial ground floor slab. Courtesy, GB Geotechnics.

concrete slab is uniformly supported by the compacted fill below the slab and the subsoil below the fill. Defects in the support can result in serious cracking. Incorrectly located reinforcement can also result in cracking. Impulse radar has been used successfully in the investigation of such defects. Figure 8.2 shows the result of an investigation by impulse radar of a large reinforced concrete slab heavily used by fork-lift trucks and subjected to high loading by stacked materials. The results of the radar survey were used for the preparation of a programme for pressure grouting to restore support to the floor slab.

8.3 SPECIAL PROBLEMS

8.3.1 Seepage of liquids through the slab

In floors seepage through the concrete slab is liable to occur:

1. when the floor is used for a 'wet' trade;
2. in balconies and external suspended accessways;
3. in the parking areas of multi-storey car parks.

8.3.1.1 Wet trades

When a floor is used for a 'wet' trade the liquid involved would be wash water from the floor, usually containing spillage from the trade carried out. The seepage would occur mainly through defective joints, but other locations to be investigated would be cracks which extend through the floor slab, alongside drainage channels, and around gulleys and manholes, and other features which penetrate the floor slab.

Needless to say, it can be very difficult, expensive and disruptive to the trade to effect a satisfactory long-term remedy.

Sometimes the spillage from the trade process is aggressive to concrete and any defects in the floor which allow the aggressive liquid to penetrate the concrete will increase the severity of the attack. The trades involved include food processing, dairies and similar. In such cases, the whole of the surface of the floor slab should be provided with a watertight, chemically resistant finish, laid to adequate gradient (say 1 in 80), discharging to properly designed floor channels. If the gradient and/or drainage is inadequate, then it would be advisable for recommendations for remedial work to include this work as well as repairing all points of seepage.

In such establishments 'ponding' is often a source of complaint, but is very difficult to avoid/eliminate.

While a gradient of 1 in 40 (25 mm in 1.00 m) should be adequate to prevent 'ponding', it may well be too steep for the safety of the workers as the surface would have to be smooth to assist in proper cleaning. A fall of 1 in 80 (25 mm in 2.00 m) should ensure effective run-off and should in itself not be a source of danger to the workers. It must be realized that even a millimetre depth of water on a floor will be seen and is often described as a 'pool of water'.

Some comments on slippery floors have been given in section 8.2.5 of this chapter.

8.3.1.2 Balconies and external suspended accessways

The repair of defective cracks and joints in what may be termed 'dry floors' have been dealt with in sections 8.2.6 and 8.2.7 respectively. The following additional comments apply when the cracks and joints have to resist the passage of water.

(a) Cracks

These cracks pass right through the slab and would probably be due to thermal contraction in the early life of the concrete, in which case they would be parallel to the main reinforcement. Cantilever balconies and external accessways are prone to this type of cracking. (see Figure 8.3.).

Balconies and accessways are usually provided with a water-proof covering such as mastic asphalt or a polyurethane-based compound. When seepage occurs, it means that the 'waterproof' membrane is damaged or deteriorated.

To make a really effective repair it is necessary to do so from the top surface of the slab and this involves removal of the covering along the line of the crack and for at least 50 mm each side. If it is not practical to repair from the top surface then the repair must be carried out from the soffit of the slab.

The crack should be cut out to a depth of not less than 25 mm and preferably 40 mm, and to a width of 15 mm, cleaned out by compressed air, the sides given a coat of primer and the repair mortar applied with a trowel. The repair mortar can be epoxy resin and fine sand as cracks of this type are normally quite static.

Other cracks through which seepage is taking place can be dealt with in the same way, but the crack-filling material should be flexible if the crack is judged to be 'live'.

In certain cases crack injection would provide a satisfactory solution; see Chapter 6, section 6.3.

Figure 8.3 View of thermal contraction cracks in soffit of balcony slab. Courtesy, Prof. B.P. Hughes, Birmingham University.

(b) Joints

The basic method of repair is similar to that described in this chapter, section 8.2.7. This may require remaking both sides of the joint for its full length and the selection of a flexible sealant which will bond well to the remade sides of the joint.

For high-class commercial property, it may be desirable to consider the installation of a proprietary mechanical jointing system, as referred to in the next section on multi-storey car parks.

As far as contraction joints and stress relief joints are concerned these would also require a flexible sealant if it is judged that movement across the joint is continuing. If no movement is taking place then a rigid or semi-rigid mortar can be used and the joint is then virtually locked.

8.4 FLOORS OF MULTI-STOREY CAR PARKS

The parking floors of multi-storey car parks should be watertight; this means that unless the concrete floor slabs are covered with a hard-wearing waterproof barrier, such as mastic asphalt or a polymer resin-based coating, the concrete slab itself has to be watertight. The reason for this requirement is that water passing through concrete becomes highly alkaline due to the high pH of the cement paste (pH about 13.5), and if this water drips onto cars serious damage is done to the paint. In winter, cars bring into the car park water from roads often containing de-icing salts. If this water penetrates into the concrete and reaches the rebars, serious corrosion will occur; this has been referred to in Chapter 3, section 3.2.6.

An investigation should consider the need for the provision of a complete waterproof layer of mastic asphalt or other suitable material such as Tretodek elastomeric polyurethane coating which has an Agrement Certificate. It is essential that the material used should be durable as a running surface. Membranes used for waterproofing bridge decks are not generally suitable as they require protection against traffic.

It is essential that repairs to cracks and joints should be carried out with great care and attention to detail. Crack injection should be considered as it is likely to provide a satisfactory solution for cracks which are static.

When sealants and mortar are used, the material should be resistant to attack by solutions of de-icing salts and petroleum oils.

Polymer-cement mortar and epoxy and polyurethane resins provide satisfactory resistance.

Movement joints are particularly difficult to deal with as movement which can only be estimated due to the many factors involved; they also have to resist moving loads, acceleration and braking of vehicles, and remain completely watertight under these arduous conditions. If a proprietary mechanical jointing system has been used, then the reason for the failure needs careful investigation with the co-operation of the suppliers before a decision for remedial work is decided upon.

If the joint has been designed 'in-house' then my experience is that serious consideration should be given to its replacement by a proprietary mechanical joint system such as Servicised LM50, or 'Waboflex SR', both marketed by Servicised Ltd. The cost of fitting such a system into an existing floor would be very high. The alternative would be to remake the joints in the 'usual way' and then provide gutters under the joints to catch the seepage which would undoubtedly occur. See Figures 8.4 and 8.5.

Figure 8.4 Defective joint in floor of multi-storey car park.

Figure 8.5 Sealed movement joint in floor of car park. Courtesy of Servicised Ltd.

PART 2 REPAIRS TO CONCRETE ROOFS

8.5 INTRODUCTION

The principal reason for repairing a concrete roof is to remedy leakage. Leakage is readily seen from the inside of the building, but it can be difficult to determine the location of the defects in the waterproof membrane and to locate the defects in the concrete slab. Unless the membrane is fully bonded to the concrete roof slab, the defects in the membrane are most unlikely to coincide with the defects in the slab.

The falls on the roof are usually formed by means of a cement/sand or proprietary screed, but experience suggests that the gradients are often inadequate resulting in 'ponding' and this is discussed later in this part.

There are many ways of constructing flat concrete roofs, the differences being in the relative positions of the base slab, the thermal insulation, and the waterproof membrane.

8.6 INVESTIGATIONS

8.6.1 Types of construction

The engineer called in to investigate alleged leakage has first to ascertain the method of construction of the roof, particularly as the 'alleged' leakage may in fact be condensation. This involves an understanding of the basic types or methods of flat roof construction, and these are described briefly below.

1. Concrete slab with no thermal insulation; waterproof layer on top of slab. This is unusual and condensation is probably inevitable.
2. Concrete slab with thermal insulation below the slab; waterproof layer on top of slab (known as 'Cold Deck' design). Vapour barrier below insulation and ventilated cavity between soffit of slab and thermal insulation. Likely design and/or construction faults in older roofs would be the omission of the vapour barrier and/or ventilated cavity.
3. Concrete slab with thermal insulation on top of the slab with waterproof layer on top of the insulation. Vapour barrier may be located on top of the slab below the thermal insulation. Known as 'warm deck' or 'inverted' or 'upside-down' design. Likely design fault in older roofs would be the omission of the vapour barrier.
4. Concrete slab; waterproof membrane on top of slab; thermal insulation on waterproof membrane; 'loading layer' of shingle on the insulation to prevent uplift. This is an 'inverted' or 'upside-down' or 'warm deck' design.

Apart from the concrete slab, the following materials are in general use.

1. Waterproof layer/membrane:
   (a) mastic asphalt or polymer-modified mastic asphalt;
   (b) built-up roofing felt;
   (c) single-ply polymeric sheet membrane.
   (d) in situ spray-applied polymers such as polyurethanes.
2. Vapour barrier:
   polyethylene sheeting, minimum 1000 gauge.
3. Thermal insulation:
   extruded polystyrene slabs or boards.

8.6.2 Inspection

The first step is to make a careful inspection of the roof, after being shown the areas of dampness on the ceiling below the roof slab. The following points should be noted.

1. The type of traffic which uses the roof. This may be occasional foot traffic for inspection and repair, or regular rather heavy pedestrian use by the occupants of the building. The use of the roof for car parking is a special case.
2. The location, size and depth of any ponding: ponding will show itself in dry weather by staining and algae growth.
3. The general condition of the roof as far as this can be seen. It should be noted that with construction methods (1), (2) and (3) in section 8.6.1, the waterproof layer/membrane is visible. However with design (4), it is underneath the 'loading' layer which means that the waterproof membrane cannot be inspected and its condition assessed until the shingle layer has been removed. This removal may cause some damage to the membrane.
4. Deterioration of the membrane can show as cracks, blisters, ripples, crazing, opening and loss of bond at joints and laps.
5. Movement joints in the concrete slab are difficult to seal and are a frequent source of leakage. Structural movement joints where there is complete separation between the adjacent concrete bays can only be effectively sealed by either a proprietary mechanical joint such as Servicised 'Waboflex SR' or Servicised 'LM 50', or by the provision of a kerb on each side of the joint as shown in Figure 8.5.
6. The junction between the roof and vertical members, such as walls and kerbs needs special attention.
7. The roof drainage, outlets and down pipes should be checked; the detailing of the membrane at outlets is often poor.
8. A thorough inspection of the underside of the roof slab is necessary. In the case of a 'cold roof' design, this would involve access to the ventilating void below the slab.

If the source of the leakage is not reasonably evident from the inspections mentioned above, then some form of leak detection procedure should be considered. There are a number of proprietary systems on the market, but my experience is that the results can be disappointing/inconclusive.

One method which I have sometimes found to give satisfactory results on roofs where the membrane is exposed is as follows. The roof is divided into a number of squares or rectangles and these formed into ponded areas (not as easy as it sounds). The water in these 'ponds' is mixed with a special dye which is colourless in normal light, but under ultra-violet light it shows as a bright blue. If staining on the underside of the roof slab is not important then other dyes can be used which are visible in ordinary light.

8.7 DIAGNOSIS

From information collected as described in the previous section it should be possible for the engineer to prepare a reasonable report including basic recommendations for remedial work. This may include upgrading the thermal insulation and the method of disposal of rainwater.

8.8 REMEDIAL WORK

8.8.1 Patch repairs

If the areas of water ingress are limited in number and extent, then patch repairs can be given serious consideration. However, great care should be taken over this as very small defects can be missed. The material used for patch repairs must be compatible with the existing membrane and must bond strongly to it. Only repair systems for which an Agrement Certificate has been issued should be used, and a list of completed work should be obtained and a few spot inspections carried out.

8.8.2 Repairs to joints

In the case of defective joints where appearance is not of great importance, e.g. covered parking areas, the fixing of gutters below the joints can provide a long-term solution at minimum cost. It is most important that water seeping through concrete roofs over parked cars should not fall onto the cars as this can cause serious damage to the paint.

8.8.3 Complete refurbishment

It may be that the most cost-effective long-term solution is to renew the existing membrane and carry out a complete refurbishment of the roof.

This should include:

1. A review of the roof gradients as many roofs have quite inadequate falls; if ponding is to be avoided, a minimum fall to outlets of 1 in 40 (25 mm in 1.00 mm) is required. In many cases this may be impossible to achieve and then some ponding must be expected and accepted by the client.

   Drainage outlets should be carefully examined as these often give trouble and may need replacement and increased in size and/or number.

Reference should be made to BS 6367: Code of Practice for drainage of roofs and paved areas, as this provides essential information and recommendations for roof drainage including calculation of run-off.

2. Careful consideration should be given to the selection of the new waterproof membrane, and whether it will be exposed or covered. For the different methods of flat-roof design, reference should be made to section 8.6.1.

   The waterproof membrane can be laid and secured in three ways:

   (a) laid 'loose' and secured with mechanical fixings to the roof deck;
   (b) fully bonded to the roof deck, as in 'warm deck-inverted roof' with thermal insulation laid on the membrane;
   (c) partially bonded to the insulation as in the 'warm deck-sandwich' roof.

Figure 8.6 'Torch-on' roofing membrane. Courtesy, Ruberoid Ltd.

The joints in the membrane must be carefully made strictly in accordance with the membrane manufacturer's instructions. The main types of membranes are listed in section 8.6.1 above. Types of membrane which are not covered by a National Standard should only be specified if they hold an Agrement Certificate.

The laying of built-up bitumen felt membranes is covered by CP 144: Roof Coverings, BS 8217: 1994: Built-up Roofing; and the laying of mastic asphalt for roofs is covered by CP 144: Part 4: Mastic Asphalt. (See Figure 8.6.)

3. The detailing of the junction between the roof and vertical surfaces such as walls and a review of joint sealing should be checked and improved as necessary. If new lead flashings are provided, they should be Code 3 (1.32 mm thick), BS 1178: 1982.
4. Thermal insulation should be upgraded if required, so as to meet the requirements of the Building Regulations, in particular the latest edition of Approved Document L – Conservation of Fuel and Power.
5. If in the refurbishment the new membrane will be exposed, then the application of a solar reflexive paint can bring a number of advantages:
   (a) reduction in degradation by ultra-violet rays and other parts of the solar spectrum;
   (b) reduction in thermal stresses as heat will be reflected from the surface and less will be absorbed.
6. Generally, the work should be carried out in accordance with BS 6229: 1982: Code of Practice for flat roofs with continuously supported coverings.

8.9 FURTHER READING

Repairs to concrete floors

British Standards Institution (1988) *Specification for Mastic Asphalt for Roads, Footways and Pavings in Building* , BS 1447: 1988.

British Standards Institution (1988) *Specification for Mastic Asphalt for Building and Civil Engineering*, BS 6925: 1988.

Building Research Establishment (1990) *The U-value of Ground Floors, Application to Building Regulations*, IP/390, 4.

Building Research Establishment (1971) Damp-proofing solid floors, *Digest*, **54**, 8.

Building Research Establishment (1993) *The U-value of Solid Ground Floors with Edge Insulation*, IP 7/93, April, 4.

Building Research Establishment (1990) *An Introduction to Infra-red Thermography for Building Surveys*, IP 7/90, July, 4.

Building Research Establishment (1979) Fill and hardcore, *Digest*, **222**, Feb., 8.
Building Research Establishment (1983) Hardcore, *Digest*, **276**, Aug., 4.
Brough, R., Malkin, F. and Harrison, R. (1979) Measurement of coefficient of friction of floors, *J. Phys. D: Appl. Phys*, **12**, 517–28.
Dow Construction Products (1988) *Floormate*, Ref. No. 4.2.1.1, July.
Harrison, R. and Malkin, F. (1983) On-site testing of shoe and floor combinations, *Ergonomics*, *26* (1) 101–8.
Sadegzadeh, M. and Kettle, R.J. (1988) Development of an accelerated abrasion test apparatus with a standardized testing procedure, *Materials and Structures*, **21**, 47–50.

Repairs to concrete roofs

British Standards Institution (1983) BS 6367: *Code of Practice for Drainage of Roofs and Paved Areas.*
British Standards Institution (1982) BS 6229: *Code of Practice for Flat Roofs with Continuously Supported Coverings.*
British Standards Institution, CP 144: *Roof Coverings*; Part 4: *Mastic asphalt.*
British Standards Institution (1994) BS 8217: *Code of Practice for Built-up Roofing.*
Building Research Establishment (1986) Flat roof design: the technical options, *Digest*, **312**, 14.
Building Research Establishment (1987) Flat roof design-thermal insulation, *Digest*, **324**, 4.
Building Research Establishment (1981) *Solar Reflective Paints*, IP. 26/81, 4.
Building Research Establishment (1979) Survey of falls in flat roofs, *BRE News*, Autumn, 2.
Building Research Establishment (1986) Wind scour of ballast on roofs, *Digest*, 311, 4.
Building Research Establishment (1992) Flat roof design: waterproof membranes, *Digest*, **372**, June, 8.
Dow Construction Products/BBA (1910) *Roofmate in Inverted Roof Concept*; Agrement Certificate No. 87, 9.
Ministry of Public Buildings and Works (1970, 1971) *Condensation in Dwellings*, Part 1: 1970, 51; Part 2: 1971, 53.

9

Repairs to concrete structures

PART 1 REPAIRS TO CONCRETE LIQUID-RETAINING STRUCTURES

9.1 INTRODUCTION

This part of Chapter 9 will deal with repairs to existing structures which have been in use for some years, and not with new structures in course of commissioning. However, when considering diagnosis and repairs reference will be made from time to time to standard requirements for new structures as these provide a guide from which departures can be seen.

The usual reason for investigating the condition of a liquid retaining structure is that leakage is known as a fact or is suspected. However, in the case of tanks holding liquids other than water, such as sewage tanks, sludge digestion tanks and slurrey tanks, there may be visible damage to the inside of the tank caused by chemical attack.

The leakage may be:

1. outward from the structure;
2. inwards into the structure;
3. some combination of (1) and (2).

9.2. INVESTIGATIONS

An investigation should follow the basic recommendations for concrete building structures as set out in Chapter 4, but with some important additions. The lines of the investigation may be summarized as follows:

1. the collection of available information about the structure, e.g., age, method of construction, the type of liquid it is required to hold (potable water, trade effluent, sewage, sludge etc.);

2. an initial inspection based on the information obtained under (1) above;
3. a visual examination of the concrete in contact with the retained liquid for possible chemical attack;
4. a preliminary assessment of the magnitude of the leakage if the location of points of leakage are visible; in certain structures infiltration from ground water and/or leakage through the roof may be occurring.

The information obtained under (1), (2), (3) and (4) above should enable the engineer to formulate proposals for a detailed inspection and investigation.

The matters which are likely to need special attention are:

1. leakage of the retained liquid;
2. infiltration of ground water and/or leakage through the roof;
3. chemical attack on the concrete by the contained liquid.

Figure 9.1 shows the results of an impulse radar survey of a large swimming pool built in the 1930s. The survey was carried out to ascertain the details of construction in view of serious cracking of the pool shell.

The survey successfully determined the thickness of the walls and floor, the location and arrangement of reinforcement and the location of voids below the floor and thus pinpointed the main causes of cracking.

9.2.1 Testing for leakage (liquid loss)

The amount of leakage which can be tolerated depends on a number of factors of which the principal are:

1. the volume lost as a percentage of the volume held;
2. the type of liquid held;
3. whether the leakage is visible (as in a water tower), or invisible (as in an underground service reservoir).

It must be kept in mind that no concrete structure will be what is known as 'bottle-tight' unless it is lined with a liquidproof membrane. In the case of water towers which are very conspicuous structures, even damp patches will show and these are often further disfigured by fungus growths. My experience is that it is prudent to provide a waterproof membrane on the inside of the tank.

With new structures, commissioning requirements accept some loss of water. The relevant Code of Practice, BS 8007: 1987: *Code of Practice for*

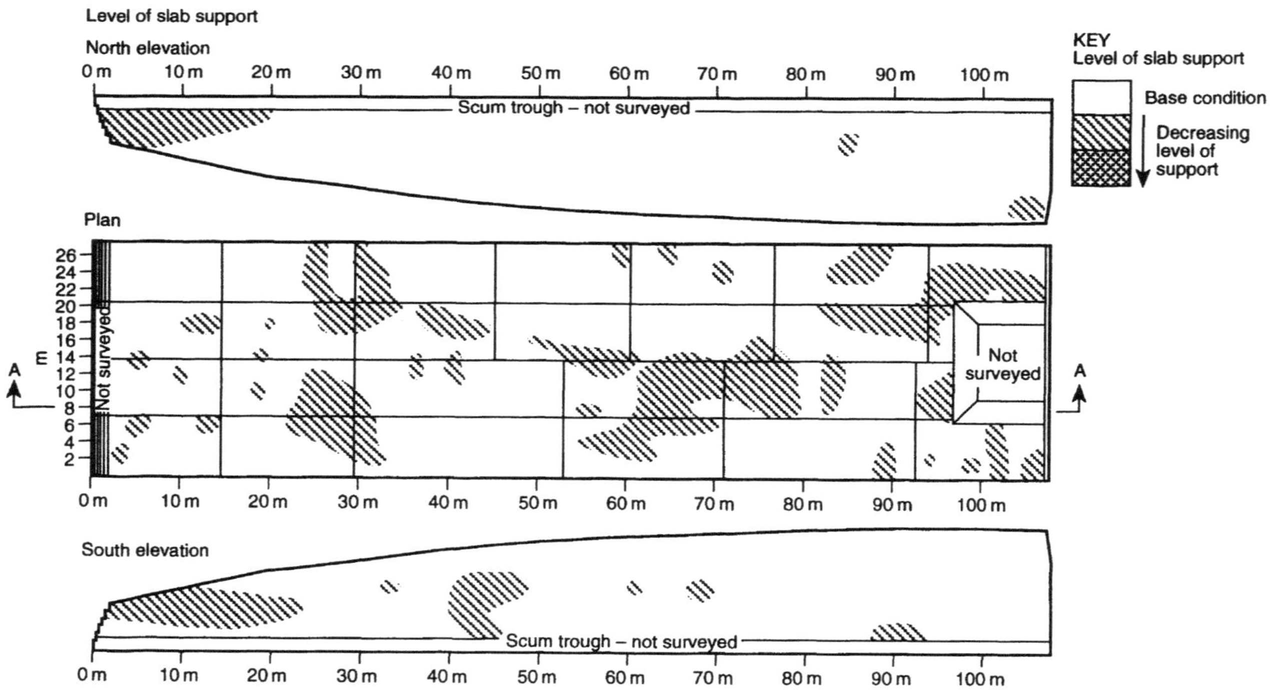

Level of slab support
North elevation
0 m
10 m
20 m
30 m
40 m
50 m
60 m
70 m
80 m
90 m
100 m
Scum trough – not surveyed
Plan
26
24
22
20
18
16
14
12
10
8
6
4
2
m
A
A
Not surveyed
Not surveyed
South elevation
KEY
Level of slab support
Base condition
Decreasing level of support

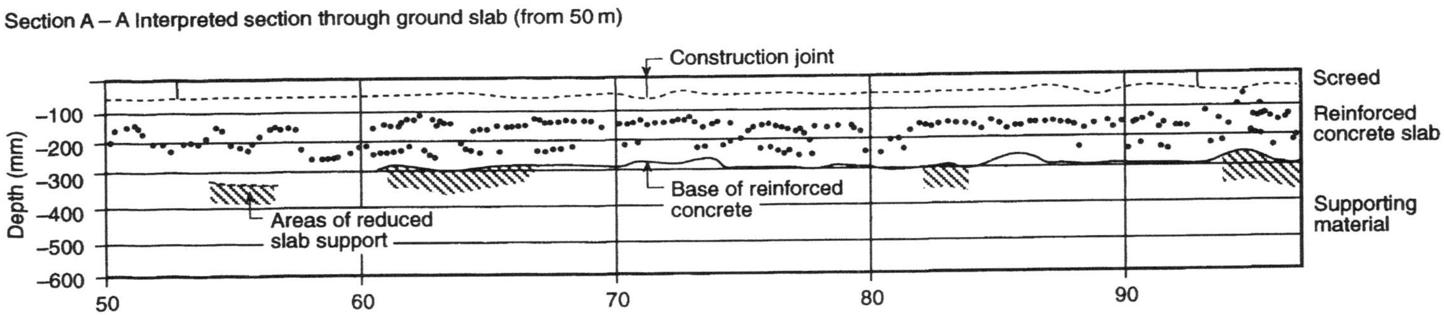

Figure 9.1 Diagram showing part assessment of swimming pool construction and condition, using impulse radar. Courtesy, GB Geotechnics.

the Design of Concrete Structures for the Retention of Aqueous Liquids, contains recommendations for assessing acceptable water loss under detailed conditions of test for new structures.

I recommend that liquid loss should be checked by means of a test as set out in the BS to which reference should be made for details. For an existing structure the test should be as set out below as far as this is practical:

1. Close all outlet valves.
2. Fill the structure with water to top liquid level, which level must be clearly marked/recorded.
3. Arrange for the measurement of evaporation during the test by means of a drum filled with water and fixed in the structure so that the water levels in structure and drum coincide. This also applies to roofed structures.
4. If the structure is open, rainfall during the test should be recorded by a rain gauge.
5. The test should continue for seven days, with the drop in water level in both structure and drum being carefully recorded at the same time each day.

The drop in water level due to leakage is the total drop, minus evaporation, plus the rainfall.

It should be noted that if the structure has been out of commission and empty for a considerable period (say, exceeding six months), then a preliminary soaking or stabilizing period of 7–21 days (depending on an assessment of the amount of drying out which has taken place) should be included in the test.

The Code recommendations for water loss in new structures are:

- 1/500 of average water depth, or
- 10mm, or
- some other specified amount.

The last alternative is realistic and applicable to existing structures.

9.2.2 Roof leakage

Structures holding potable water are usually provided with a roof and to reduce contamination this should be watertight, particularly if, as is often the case, the structures are partly buried and the roof covered with earth and grass.

In the case of existing structures, it is not unusual to find some leakage through the roof. If this consists of damp/wet areas where the amount of seepage is too small to be measured and it has not resulted in corrosion of the reinforcement, then it is likely that the cost of necessary work to completely eliminate this moisture penetration would not be justified.

Where the leakage is more serious, then remedial work would be necessary. The Code contains recommendations for testing new flat concrete roofs for watertightness. This consists of flooding the roof to a depth of 25 mm and leaving this for 24 hours; if there are no visible leaks or damp patches, then the roof can be considered as satisfactory. Should it be difficult to flood the roof for 24 hours then, as an alternative, water should be arranged to flow over the roof surface for six hours, and the soffit inspected for seepage and damp patches.

9.2.3 Location of leakage

The detection of places where leakage is taking place is comparatively easy with above-ground elevated structures such as water towers. With structures partially or completely under ground considerable difficulty exists. The three most likely locations for leakage are through joints and cracks, and where pipes pass through the walls below top liquid level and through the floor. It is usual practice to provide such pipes with a puddle flange, but this will only be effective if the concrete has been thoroughly compacted around the pipe.

All joints should be sealed and the condition of the sealant (if one is provided) should be carefully examined. Joints without a sealant are suspect from the point of view of leakage.

It may be difficult to decide whether a crack is leaking by merely looking at it. It is necessary to try to decide on the type of crack, e.g. thermal, contraction, drying shrinkage or flexural. Its location, width and length are also relevant.

Leakage through floor slabs on the ground is difficult to detect unless there is an accessible system of underdrains. If there is, then the flow through the underdrains would increase with about 1 m head of water on the floor slab if the floor was not watertight. Even when it is reasonably clear that leakage is taking place through the floor slab, there remains the need to locate the points of leakage and to carry out effective repairs. This presents many problems.

9.2.4 Infiltration

With the structure empty, significant infiltration should be visible, but slight seepage will be difficult to detect until the surface of the floor and inside surface of walls and soffit of the roof slab have dried out. Even a small amount of infiltration would be unacceptable in potable water tanks if the water stored was after purification.

9.2.5 Corrosion of reinforcement

The whole of the inside surface (and the outside if visible) should be carefully examined for signs of reinforcement corrosion which usually shows as rust staining with cracking and spalling of the concrete.

When steel is corroded and converted into rusts (oxides of iron), the volume of rust is three to five times the volume of the original metal and the expansion can crack and spall the concrete.

There are two types of corrosion, namely, general corrosion and local corrosion. The general corrosion is more likely to result in spalling of the concrete; local corrosion can sometimes appear in the form of pitting which may be more serious to the residual strength of the reinforcement owing to the reduction in the sectional area of the bars at the 'pits'. These pits may penetrate the rebar by more than 50% of the bar diameter. Localized corrosion is more likely to result from the presence of chloride ions in the concrete in contact with the steel, but it does not necessarily cause spalling of the concrete. Therefore even a careful visual examination of the concrete may not detect localized/pitting corrosion.

Figure 9.2 Deteriorated concrete slurry tank due to reinforcement corrosion. Courtesy, Cement Gun Co. UK Ltd.

A cover-meter survey will show areas where the cover to the rebars is less than that normally considered as providing adequate protection, which is 40 mm. The UK Code of Practice, BS 8007: 1987, recommends that this depth of cover should be increased if the concrete is in contact with aggressive liquids (either in the tank or external in the sub-soil). See Figure 9.2 showing deterioration of an rc slurry tank.

A half-cell survey (see Chapter 4, section 4.5.5) will indicate locations where active corrosion of the rebars is likely to be taking place, even though there are no visible signs of this activity.

For above ground structures, a check should be made for depth of carbonation for the reasons given in Chapter 3, section 3.2.5; see also Chapter 4, sections 4.4.2 and 4.5.2. Carbonation itself does not cause a reduction in the strength and water resistance of the concrete; it is its effect on the corrosion of steel reinforcement that is important.

9.2.6 Chemical attack on the concrete

Chemical attack can occur to concrete from the liquid stored in the structure, and from ground water. Chapter 3, section 3.5 gives information on the effect of a number of chemicals in general use on concrete. As far as water-supply structures are concerned, chemical attack is more likely to

Figure 9.3 Serious attack by acidic trade effluent on concrete sewer.

be found in contact tanks in water treatment works, then in reservoirs. However, reservoirs, pipelines and aquaducts holding and conveying water of low pH derived from peaty areas, such as upland gathering grounds, may suffer attack. The pH of such water can fall as low as 3.5 and although this is due to organic acids, quite severe etching of the concrete can result. Figure 3.5 shows damaged concrete water channels carrying soft moorland water. See also, Chapter 3, section 3.5.9. Figure 9.3 shows the damage done to concrete sewer pipes by an acidic trade effluent.

9.2.7 Hydrogen sulphide corrosion

Tanks in sewage treatment works, sludge digestion tanks and sumps of pumping stations, can under certain special conditions, be subjected to severe attack above top water level. This has been briefly referred to in Chapter 3, section 3.5.11, and is known as hydrogen sulphide corrosion. There have been reports of this type of attack in the UK, and I have investigated a number of cases and feel that brief information on where conditions conducive to this type of attack are likely to be found, would be helpful to readers.

1. Long lengths of sewer with a flat gradient and non-self-cleansing velocity and sumps in pumping stations receiving discharge from such sewers, or where the retention period in the sump is prolonged.
2. Sewers receiving trade effluent which contains high concentrations of sulphide or organic sulphur compounds; for example, brewery waste. High temperature in the sewage will accelerate septicity and the formation of hydrogen sulphide (H_2S) gas.
3. Pumping stations sumps with long retention periods, and the discharge points of pumping mains from such sumps.
4. Sludge digestion tanks: theoretically, when properly controlled these tanks should not suffer this type of attack, but it does happen sometimes.

It is emphasized that maximum safety precautions must be taken when inspecting sewers, manholes, sumps, covered tanks etc. Hydrogen sulphide gas and other gases generated in septic sewage are poisonous. Adequate ventilation prior to entry, at least two men present at the surface access, the use of a proper harness by those entering; and the air inside the sewer etc. should be tested for oxygen, carbon dioxide and hydrogen sulphide; concentrations of these gases should be similar to those in the open air.

The requirements of the Health and Safety Executive must also be followed.

As stated in Chapter 3, the H_2S from the septic sewage rises up into the very humid airspace above top sewage level where it is converted to sulphuric acid by aerobic bacteria. Sulphuric acid (H_2SO_4) is one of the most aggressive acids to Portland cement concrete and can cause destruction to the structures mentioned above. I have seen an rc column in a pumping station sump almost completely destroyed in this way.

9.2 8 Sludge digestion tanks

Sludge digesters are often included in sewage treatment works. They are usually circular on plan with domed roofs. The walls and floor are inforced concrete (sometimes the walls are post-tensioned) and the roof is heavily galvanized steel or reinforced concrete or shotcrete. Heating and mixing equipment is provided inside the tank. The sludge is usually maintained at 25 °C–30 °C, and the optimum pH is 7.0 to 7.8. Under these conditions the sludge is either neutral or alkaline and a considerable volume of gas is given off by the digesting sludge; the gas consists of about 70% methane and 25% carbon dioxide, the remaining 5% consists of hydrogen sulphide and other gases.

Unfortunately if strict control breaks down or if the characteristics of the sludge change, the pH can fall and the amount of H_2S can increase considerably. This can result in conditions in the digester becoming suitable for the hydrogen sulphide-sulphuric acid cycle to occur in the airspace above top sludge level.

9.2.9 Damage to concrete by freeze–thaw

This type of damage is only likely to occur to structures in very exposed positions in temperate or colder climates. It shows as weakening of the exposed surface of the concrete due to disintegration of the cement matrix, giving the appearance of fairly shallow spalling. See Figure 3.1 in Chapter 3.

9.2.10 Alkali–silica reaction

Information on this type of attack on concrete structures has been given in Chapter 3, section 3.5.13, in Chapter 4, section 4.6.3.5, and in Chapter 6, section 6.19.7.

I know of only one definitely established case of alkali–silica reaction in a liquid retaining structure in the UK (except a dam in Jersey).

The three basic requirements for the initiation of alkali–silica reaction are:

1. Sufficient amount of caustic alkalis in the concrete
2. The presence of reactive silica in the aggregates
3. Adequate supply of moisture.

In an existing liquid retaining structure, very little can be done to mitigate the effect of any of the above three conditions. Therefore, in the event of alkali–silica reaction being confirmed, a decision would have to be taken either to demolish the members affected and replace them or to leave matters as they were on the assumption that the strength of the members will not be significantly reduced.

9.2.11 Sampling and testing

This subject has been discussed in Chapter 4, section 4.4 and the following comments relate specifically to liquid retaining structures.

It is generally advisable for samples of concrete to be tested for sulphate content, particularly concrete below ground level in a clay soil. If the sulphate content of the concrete exceeds 4% by mass of the cement, then further investigations could be justified as this would indicate either sulphate in the aggregates or the diffusion into the concrete of sulphate bearing ground water. For the effect of sulphates on Portland cement concrete see Chapter 3, section 3.5.5.

For structures below ground level or under embankments, checks for depth of carbonation are not necessary.

Testing for chlorides is generally desirable.

It may be necessary to check the general quality of the concrete in terms of cement content, approximate grading of aggregates and standard of compaction. Also, tests to confirm or reject the existence of alkali–silica reaction may be advisable if the crack pattern indicates this possibility; see Chapter 3, section 3.5.13.

If there are signs of chemical attack on the liquid-retaining surface of the concrete, then chemical analysis of the retained liquid is desirable. Information should be obtained on whether the chemical composition of the liquid changes with time. The pH of the liquid is important. It should be noted that the pH scale is logarithmic, so that the hydrogen ion concentration of a liquid with a pH of 4.0 is 1000 times that of a liquid with a pH of 7.0. The pH and chemical composition of potable (drinking) water varies. Most drinking water in the UK falls into the following three categories:

1. Soft, slightly acidic water, low in alkalinity and total dissolved solids, with a pH in the range 5.0 to 6.5. This is a typical water from upland gathering grounds, moorland areas etc.

2. Water with a pH between 6.5 and 7.5, medium alkalinity and total dissolved solids (tds)
3. Water with pH between 7.5 and 8.5, higher alkalinity and tds.

The pH can be determined approximately with indicator papers and indicator solutions, and a comparator. For more accurate determination a pH meter should be used.

Water of category (1) above is likely to etch the concrete; and over a long period the etching can be severe. (See Figure 9.4.)

Figure 9.4 Repair of concrete slurry tank with reinforced gunite. Courtesy, Cement Gun Co. UK Ltd.

Contact tanks in water treatment works where ferrous sulphate is used can suffer quite severe attack and protective coatings may be required.

The damage caused by hydrogen sulphide corrosion of the concrete is generally very severe, (see Figure 3.3, page 56) and is typically above the water line in sumps and sewers and the upper parts of walls and soffit of roof slabs in sludge digestion tanks. As the cost of remedial work is likely to be very high, full information on the characteristics of the sewage and digesting sludge should be obtained.

Figure 9.5 Deteriorated sealant in joint in wall of water reservoir. Courtesy, Colebrand Ltd and Thames Water Authority.

9.2.12 Impulse radar survey

An Impulse radar survey can be very useful when serious cracking has been detected. The survey can locate differences in the density of the concrete, location of rebars, differences in support provided by the subsoil.

Figure 9.1 shows part of a radar survey of a large open-air swimming pool.

9.3 DIAGNOSIS

A practical and sensible diagnosis of the causes of faults and deterioration in liquid retaining structures requires considerable experience.

I suggest the listing of all faults and deterioration with an opinion on the importance of each in relation to the efficient operation of the structure and an estimate of the likely cost of remedial work. The client in consultation with the engineer could then decide what work should be carried out.

The presence of structural cracks – that is, cracks which indicate a reduced factor of safety for the member(s) in question – would require special consideration and may result in a decision to take the structure out of use.

Where there has been chemical attack by the contained liquid, or caused by the contained liquid – as in hydrogen sulphide corrosion above the top liquid level in the tank, then consideration should be given to reducing or eliminating the cause of attack. A decision on whether such steps be taken would depend largely on the degree of attack and the estimated time the attack has been taking place and the practical feasibility of treating the liquid to reduce its corrosive characteristics. In the case of hydrogen sulphide–sulphuric acid attack, remedial action is likely to be complex and may require changes to the operation of the sewerage system.

9.4 REPAIRS AND REMEDIAL WORK

9.4.1 Work required to remedy/reduce leakage/infiltration

As previously stated, major water loss is more likely to occur through joints and cracks than through areas of honeycombed concrete. A gunite repair is shown in Figure 9.4.

For tanks holding potable water, all repair materials must be non-toxic, non-tainting, should not support bacterial growth and must not adversely affect the quality of the water.

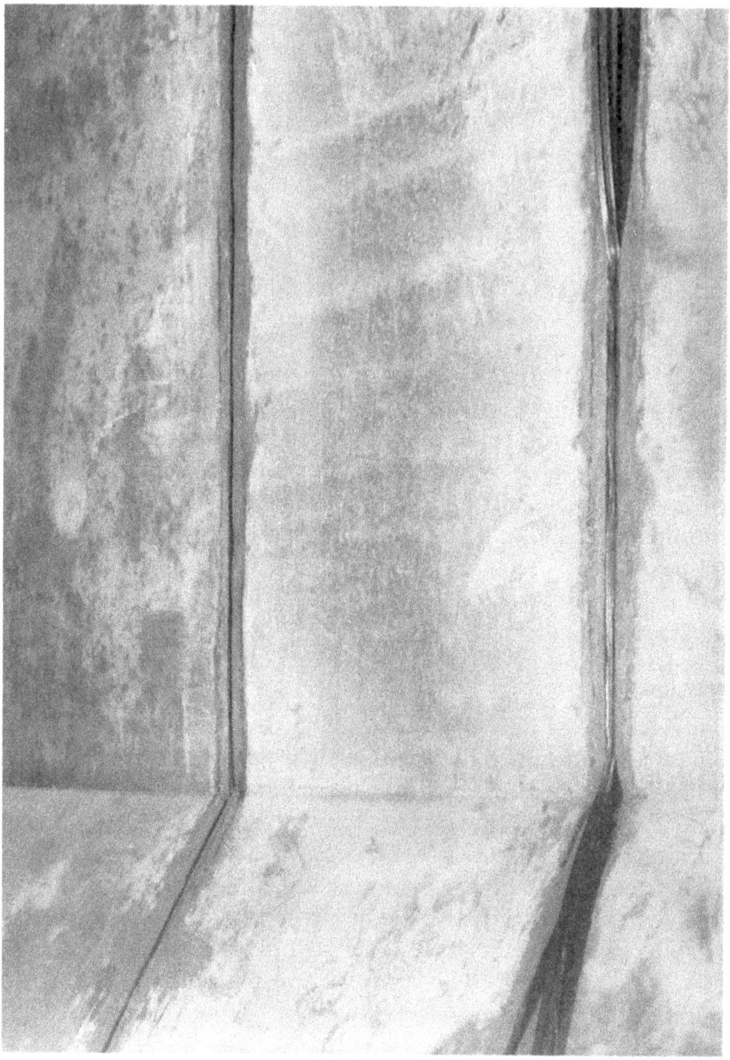

Figure 9.6 Joints in wall of water reservoir being resealed with Neoferma-EPDM gaskets. Courtesy, Colebrand Ltd and Thames Water Authority.

9.4.2 Remedial work to joints

It is usual for joints to be provided with a sealing groove sealed with a flexible durable sealant and with a water bar. I have found cases where the sealant has been omitted. Joints extend for the full thickness of the wall, floor or roof and therefore must be watertight.

If a sealing groove has not been provided, then it may be advisable, depending on the type of joint, to cut a groove and seal it on the water face. An alternative is to use crack injection as described in the next section.

If a sealing groove has been provided, then the old deteriorated sealant must be removed, the joint carefully cleaned out and a new sealant inserted. The new sealant must be durable under the environment in which it has to operate. (See Figures 9.5 and 9.6.)

The sealant used on the outside of an elevated tank, in an exposed position where freeze–thaw conditions exist in the winter must be able to function effectively under such conditions.

Sealants used in tanks holding sewage and trade effluent and sludge digestion tanks operate in particularly onerous conditions. These 'conditions' should be carefully assessed and full information given to the sealant suppliers.

Sealants are normally used with a primer, and this should be suitable for application to damp concrete. It is unlikely that the surface of the joint will be effectively dried.

Some general information on sealant types has been given in Chapter 2, section 2.13.

9.4.3 Remedial work to cracks

Cracks, unlike joints, do not necessarily extend right through the walls, floor or roof. Therefore the need to repair them will normally depend on:

1. whether they are associated with corrosion of reinforcement revealed by rust stains and/or spalling, or indicated by a half-cell survey;
2. whether they extend through the structure and are a source of liquid loss when the tank is full or infiltration when the tank is empty.

It is my experience that cracks not associated with rebar corrosion nor leakage do not need repair and can be left alone, unless they are assessed as denoting structural weakness.

When decided that repair is necessary, the method of repair will depend on type of crack and whether further movement across the crack is anticipated, and if so the estimated magnitude of the movement.

For cracks not associated with further movement, the following method of repair is suggested:

1. The laitence should be removed from the concrete for a distance of 75 mm each side of the crack. This can be done by power operated wire brushes, light grit blasting, or high-velocity water jets. All grit and dust must be removed from the prepared surface.

2. Brush into the line of the crack and onto the prepared area three coats of suitably formulated low viscosity slightly flexible epoxy resin, preceded by the application of a suitable primer.

An alternative is to use crack injection, but it would be prudent to carry out a trial first, unless the outside of the wall was accessible so that the outside of the joint could be sealed. Unless this is done an unlimited amount of high-priced resin may be pumped in and lost without effectively sealing the joint. The resin used should be slightly flexible and be of low viscosity.

For details of crack injection technique, see Chapter 6, section 6.3.

9.4.4 Remedial work to spalled concrete

For spalled concrete resulting from rebar corrosion, the work should be carried out as follows.

1. The concrete should be removed to expose the corroded rebars followed by removal of all rust from the rebars.
2. Grout the rebars and the surface of prepared concrete with a cement/SBR grout (about 25 litres of SBR to 50 kg OPC).

Figure 9.7 Open-air swimming pool being finished with chlorinated rubber coating after completion of repairs to joints, cracks and spalled concrete. Courtesy, Colebrand Ltd.

3. Apply the repair mortar, consisting of cement and prepared clean graded sand, gauged with sufficient water and SBR to give a stiff but workable mix. Suitable mix proportions would be:

   50 kg cement (OPC or RHPC);
   150 kg clean sharp sand;
   10 litres of SBR to 50 kg cement.

4. These patch repairs should be cured in a manner determined by site conditions.

See Figure 9.7 showing coating of walls and floor of an open air swimming pool after extensive remedial work.

9.4.5 Honeycombed concrete

The repair of honeycombed concrete can be carried out either by a 'surface' repair, i.e. by cutting out the affected concrete and replacing it with new concrete or mortar depending on the location and depth to which the honeycombing has extended.

The repair of honeycombed concrete in beams and the soffit of suspended slabs would have to be done by either hand applied mortar, or by pneumatically applied mortar.

The main problem is to ensure that the new mix is well bonded to the old concrete and is thoroughly compacted. It may be necessary to provide temporary support to the member under repair, depending on the magnitude of the honeycombed area and its location. All honeycombed and sub-standard concrete should be cut away and this can be done by percussion tools or high-velocity water jets. The former produces noise and dust but, with the latter, provision must be made for the disposal of the water. Sufficient concrete should be cut away to allow for placing and compacting the new concrete. One method is shown in Figure 5.3. Placing and compacting is usually difficult and therefore consideration should be given to the use of a superplasticizer to provide a collapse slump with a low w/c ratio. After removal of the formwork, the new concrete should be cured as recommended in Chapter 5, section 5.3.5.

If the honeycombing is virtually in the concrete cover to the rebars, then the repair can in many cases be made with a high-quality cement–sand–SBR mortar. If possible, pre-bagged materials should be used, with mix proportions of one part OPC or RHPC to three parts clean graded sand and 10 litres of SBR to 50 kg cement, with sufficient gauging water to give a workable mix. These repaired areas should be cured as recommended in Chapter 5, section 5.3.5.

It is not suggested that the repaired member will necessarily have the same strength as the original was intended to have. However, if allowance is made for redistribution of stress between the old and new concrete and the reinforcement, the small reduction in strength should be insignificant.

Both the new concrete and the mortar will be subjected to drying shrinkage and a perimeter crack may form. It is therefore advisable to wire brush a band about 75 mm wide around the repaired area to remove laitence and to apply two brush coats of cement/SBR grout, as late in the repair period as practicable.

Where the honeycombing is located towards the centre of the cross section of the member, pressure grouting using a proprietary cement-based non-shrink grout may be the better solution. This work should only be entrusted to an experienced specialist firm with a proven record of satisfactory performance.

Sometimes the honeycombing is not visible and is detected by a UPV or impulse radar survey. To deal effectively with such a situation requires that either an appreciable amount of sound concrete has to be cut away in order to reach the honeycombed area, or the repair has to be carried out by pressure grouting. This requires special experience and the selection of suitable cement-based grout. The required characteristics of the grout would depend mainly on site conditions, but would include low shrinkage, no segregation, easy to pump and controlled setting time.

The result of the pressure grouting on filling the voids in the honeycombed concrete should be checked by UPV or impulse radar, rather than by cutting cores.

9.4.6 Remedial work required by hydrogen sulphide corrosion

9.4.6.1 Introduction

The damage to concrete caused by this form of attack can be very considerable unless it is detected in the very early stages and effective remedial measures put in hand.

Practical steps to rectify the situation include:

1. Changes in the operation of the sewerage system where the damage has occurred. This is a long-term aim and, strictly, is not a repair. It would include such matters as injection of oxygen or compressed air into pumping mains.
2. Chemical treatment of the sewage/sludge so that H_2S evolution is drastically reduced; hydrogen sulphide corrosion cannot occur unless H_2S is present in the air space above the sewage. Chlorine will

react with sulphides present in the septic sewage and thus prevent the formation of H_2S.
3. Installation of effective and continuous ventilation of the air space above the sewage so as to remove the H_2S as it rises above sewage level.
4. Repair the damaged concrete as briefly described below or replace the unit if damage is too serious to allow effective repair. Repairs should include the use of barrier coats to protect the new concrete.

It is emphasized that repairs are unlikely to be durable unless some steps are taken as suggested in (1) to (3) above.

Investigations and testing should include an assessment of the depth to which the sulphuric acid has penetrated into the concrete. The effect of H_2SO_4 attack on the concrete is to convert the hydrated cement paste into a compound which has the consistency and strength of a lime mortar.

9.4.6.2 *Repair methods*

Reference should be made to section 9.2.7 for the absolute necessity to take maximum safety precautions before entry into any enclosed part of the sewerage system.

Having established the probable depth of acid attack on the concrete during the investigations, all defective/acid impregnated concrete must be cut away. Concrete which has a pH below 11.5 should be removed. Due to the amount and depth of concrete to be removed it may well be necessary to use concrete for the repair rather than mortar. A suggested mix is:

> 350 kg/m^3 of OPC; 10 mm graded aggregate (high-quality limestone if available) and clean concreting sand; free w/c ratio not to exceed 0.4 and a selected plasticizer to ensure adequate workability for maximum compaction.

For repair mortar, the mix should be 1:3.5 by mass OPC to clean graded sand; both cement and sand should be pre-packed and water and a plasticizer added on site.

The use of concrete will require the use of formwork and the time taken to fix this may preclude the application of a bonding coat to the prepared concrete as the new concrete must be placed within 20–30 minutes of the bonding coat.

A bonding coat of OPC and SBR should be used for mortar repairs. I favour the application of an epoxy resin barrier coat over the surface of the concrete, extending upwards from 1.0 m below mean liquid level.

The concrete must be properly prepared to receive the epoxy primer, followed by at least two coats of the resin.

9.5 REPAIRS TO ROOF SLABS

Concrete roof slabs must be watertight for the following reasons:

1. leakage will inevitably result, in the long term, in the corrosion of the reinforcement;
2. seepage through the concrete will be highly alkaline and this may adversely affect the liquid stored in the tank;
3. if the tank holds potable water, serious contamination of the water can occur.

It is unrealistic to assume that a roof slab will be completely watertight unless it is provided with a durable water-resistant membrane. If investigations show that the roof is leaking then this means that the membrane is defective (assuming one has been provided, which is not necessarily the case in old structures).

In most cases, the practical solution is to provide a new membrane. The membrane can be either of sheet material or an in situ compound. The membrane, whether sheet or in situ, must be puncture-resistant and possess high strength to resist normal building movement, and must bond well to the concrete. A typical high-quality sheet material is Bituthene DW, which is a laminate of high-density polyethylene (HDPE) with a rubber bitumen adhesive.

A well-known and successful in situ compound is Tretodek, which is a two-pack pigmented polyurethane.

Of great importance is the detailing at the edges of the roof and around any features which penetrate the roof slab, such as manholes, ventilators, pipes etc.

It is most likely that some cracking and spalling of the concrete has resulted from the leakage and this should be repaired in accordance with good practice as described in section 9.4.4.

If corrosion of reinforcement is assessed as being due to the presence of chlorides in the concrete, then consideration would have to be given to the installation of a system of cathodic protection or to a method of chloride removal.

These methods are briefly discussed in section 9.6.1 and 9.6.2 below.

9.6 SPECIAL REPAIR METHODS

9.6.1 Cathodic protection

Cathodic protection (cp) has been described in Chapter 6, section 6.6.

The main reason for using cp in the repair of reinforced concrete is to provide permanent protection against the corrosion of reinforcement by chloride ions in the concrete. The cost of a cp installation is expensive but much less than a programme of regular concrete repair throughout the lifetime of the structure.

9.6.2 Chloride extraction

This comparatively new system can be considered as a possible alternative to cathodic protection in as much that it is intended to provide permanent protection to the rebars by the removal/ extraction of chlorides from the concrete. Experience of this system in the UK is limited, and it is briefly described in Chapter 6, section 6.7.3.

The system will extract the chlorides and thus reduce the chloride ion concentration from the exposed surface of the concrete down to the rebars and, optimistically, the concrete between the rebars. It is intended to operate in circumstances where the chlorides have diffused into the concrete from the exposed surface, as for example in deck slab of a highway bridge. To be effective, the concentration of chloride in the concrete surrounding the rebars should not exceed the maximum concentration as laid down in BS 5328, Part 1, Table 4.

Assuming the system works as intended, this would only provide a permanent solution if after the completion of the extraction process no further diffusion of chlorides into the concrete takes place.

9.6.3 Re-alkalization of concrete

This is a proprietary system for dealing with the effects of carbonation of concrete when this penetrates down to the reinforcement. The carbonation of concrete has been dealt with in some detail in Chapter 3, section 3.2.5; re-alkalization of concrete is described briefly in Chapter 6, section 6.7.2.

Carbonation of concrete is only likely to occur in concrete exposed to the air and so would only be a potential threat to the rebars in the part(s) of the liquid retaining structure above ground level.

Re-alkalization is intended to restore the pH of the concrete in contact with the rebars to a pH of 10.0 or above. Corroded rebars and any consequent damage to concrete (cracks and spalling) would have to be repaired in the usual way.

PART 2 REPAIRS TO CONCRETE WATER-EXCLUDING STRUCTURES

9.7 INTRODUCTION

For the purpose of this part of the chapter, water-excluding structures are basements used for various purposes and the retaining walls to subways (pedestrian underpasses).

As with liquid-retaining structures, the usual reason for an engineer to be asked to make an inspection of a basement is because of the infiltration of ground water; and/or 'damp patches' on walls and/or ceilings.

Associated with infiltration is the probability of corrosion of reinforcement, resulting in cracking and spalling of the concrete cover.

It must be accepted that no concrete structure will be what is known as 'bottle-tight', unless it is provided with a water-resistance membrane which to achieve the best results should be on the water side of the roof, walls or floor.

With the vast majority of basements, it is not possible to have access to the outside of the walls which means that the principle of sealing leaks on the water face cannot be applied.

This basic difficulty results in firms which specialize in this work being somewhat reluctant to give guarantees of complete success without qualification.

In the UK two relevant publications on waterproofing basements are:

1. BS 8102: *Code of Practice for the Protection of Structures against Water from the Ground*, 1990.
2. Construction Industry Research & Information Assoc (CIRIA) (1995) *Water-Resistant Basement Construction – A Guide*, Report No. 139.

The publications listed above set out clearly the level of protection required to suit the use to which the basement is put.

Concern has been expressed in technical journals at the fact that the water table in the London area is rising. In the areas affected, this may result in an increase in existing ground water pressure against deep basements while basements previously above ground water level may become subject to ground water pressure.

It will be seen from Table 9.1 that to achieve a performance level of a 'totally dry environment', mechanical ventilation including humidity control is likely to be required. This is a matter of major importance when dealing with a change of use in an existing basement from say

Table 9.1 Guide of level of protection to suit basement use

Grade	*Basement usage*	*Performance level*	*Form of construction (see Figure 6 of Code)*	*Comment*
1	Car parking; plant rooms (excluding electrical equipment); workshops	Some seepage and damp patches tolerable	Type B. Reinforced concrete design in accordance with BS 8110	Groundwater should be checked for chemicals which may have a deleterious effect on the structure or internal finishes
2	Workshops and plant rooms requiring drier environment; retail storage areas	No water penetration but moisture vapour tolerable	Type A. Type B. Reinforced concrete design in accordance with BS 8007	Careful supervision of all stages of construction is necessary. Membranes can be applied in multi-layers with well lapped joints
3	Ventilated residential and working areas including offices, restaurants etc., leisure centres	Dry environment	Type A. Type B. With reinforced concrete design to BS 8007. Type C. With wall and floor cavity and DPM	As grade 2
4	Archives and stores requiring controlled environment	Totally dry environment	Type A. Type B. With reinforced concrete design to BS 8007 plus a vapour proof membrane.	As grade 2
			Type C. With ventilated wall cavity with vapour barrier to inner skin and floor cavity with DPM	As grade 1

general storage (Grade 2) to an art gallery or storage of documents (Grade 4).

9.8 TRACING INFILTRATION AND DAMPNESS/CONDENSATION

9.8.1 Infiltration

In the case of basements the location of the infiltration points are clear when the top surface of floor and inside surface of the walls are open for inspection. When the walls and floor are finished with a bonded finish such as rendering it is reasonable to assume that the defect in the concrete is more or less in line with the wet area. However, when the finish is not fully bonded to the substrate, e.g. a 'dry-lining', the source of infiltration through the concrete may be some distance from where it shows on the wall finish.

The vulnerable parts of the walls and floor to infiltration are the joints, cracks which penetrate through the walls/floor, and honeycombed concrete.

As stated above, the repair may have to be executed on the inside face of the walls and/or floor, against the pressure of the ground water.

Ground water levels in relation to the level of the basement floor are therefore relevant and information on these levels and their variation are important when considering remedial work.

9.8.2 Dampness/condensation

Damp patches on walls and ceilings may be due to water slowly penetrating from outside, or to condensation, or to some combination of both. This makes correct diagnosis very difficult and considerable experience is required.

Where a basement is shallow, the floor is above the water table, and the walls partly above ground level, rising damp can occur in the floor and lower part of the walls due to moisture being drawn upwards from damp sub-soil and slowly penetrating through the concrete.

If the floor/walls have an impervious covering then there is likelihood that wet patches will form behind the impervious covering. This is why it is essential for floors to be provided with waterproof membrane when first constructed. The provision of a new membrane in an existing floor is difficult as the most satisfactory position for it is below the floor slab.

I would add that in my experience, **rising** damp in concrete walls is unusual, but if the walls are in contact with a damp sub-soil, then moisture may slowly penetrate through and may rise upwards for a limited distance.

Dampness originating from an external source is likely to show as isolated damp patches which vary in size according to the weather; they increase in size after prolonged rain and tend to disappear during long periods of dry weather.

Diagnosis is often made on the basis of 'moisture meter' readings. This is an electrical resistivity meter which is pressed against the wall surface. *BRE Digest*, 245, June 1981, states:

> Measurements of surface moisture are in themselves no indication that a genuine rising damp problem exists [...] Electrical meters commonly used are responsive to both the amount of moisture present and to the salt concentration in the surface layers and cannot distinguish between the two.

The recommended method for determination of whether rising damp is present or whether the dampness is due to condensation, is to take samples of the mortar and substrate and measure the water content and hygroscopity. Walls may contain considerable quantities of hygroscopic salts. The laboratory measurements will show whether the wall could have absorbed from the air the amount of water found in the samples.

There may be other sources of water such as a leaking water pipe, or waste pipe from a sink or bathroom fitting.

On the other hand, condensation can arise from design of the building and the way it is used. Basements tend to be cool and badly ventilated and when the temperature of the air in the basement reaches the 'dew-point' condensation will occur on cold surfaces. The 'dew-point' is the temperature of the air when the air is saturated with water vapour. Relative humidity (RH) can be considered as the amount of water vapour present expressed as a percentage of the amount that would saturate it at the same temperature.

The relationship between dry bulb air temperature, relative humidity, moisture content and vapour pressure is usually shown in the form of a diagram, known as a psychrometric chart.

For example, with a dry bulb temperature of 15°C and RH of 60%, the air would contain about 6.4 gms of moisture per kg. With the same temperature and a RH of 90%, the air would contain about 9.6 gms of moisture per kg. With a drop in temperature to 13 °C, the RH would be 100% (saturated). Such air in contact with a colder surface (basement wall) would deposit moisture in the form of condensation.

Reference can be made to a publication by Ministry of Public Building and Works, *Condensation in Dwellings* (1970), which deals with the subject very clearly, and to BS 5250: 1989, *Code of Practice for Control of Condensation in Buildings*.

From the above it can be seen that by raising the air temperature, condensation can be eliminated.

Associated with damp areas on walls, including when this is due to condensation, is mould growth. Remedial work to eliminate condensation will certainly reduce the likelihood of mould growth continuing, but will not necessarily stop it. It is therefore advisable to adopt the following procedure:

1. Remove all existing mould and sterilize the wall surface with at least two coats of fungicide.
2. Redecorate using a good-quality fungicidal paint.
3. If possible, take steps to reduce the relative humidity to below a level appropriate for the wall surface material.

For example, a RH of 80% with brick or painted surfaces mould growth may not develop.

9.9 REPAIR METHODS

9.9.1 General considerations: basements

The method of repair and the possibility of its success will depend on a number of factors of which the principal are:

1. The Grade of basement, i.e. the use to which it will be put. Table 9.1, which is reproduced by permission of the British Standards Institution, shows the standard of protection against water/moisture penetration needed for four Grades of use, and the type of construction needed to meet each Grade.
2. The construction of basements in older buildings seldom meet the recommendations given in BS 8002 and in the CIRIA Report which is based on the Code. This has an important bearing on the methods of repair and materials used, and even on whether it is practical to achieve the standard of dryness required for a particular use.
3. The ground water pressure and its variation and its chemical characteristics.

It will be seen that there are basically three types of construction which should achieve the various standards of dryness required by basement use. While these recommendations refer to new structures, they are an essential guide to an engineer called upon to recommend remedial work to a leaking basement or to upgrade an existing basement for change of use.

The British Standard 8102 was published in 1990 and the CIRIA Guide in 1995, so that basements constructed prior to 1990 may not meet the present requirements, and the older the basement is the greater the divergence is likely to be.

To carry out remedial work to an existing underground car park so that it meets the standard (performance level) of a Grade 1 basement should not present undue difficulty. To upgrade an existing basement to meet the performance levels for Grade 3 or Grade 4 use could be very difficult.

My experience is that the engineer should make the position quite clear to his client. It is also essential that when discussing the proposed work with specialist contractors prior to inviting tenders, the standard of dryness required should be made perfectly clear in writing. All contractors tendering should be asked to sign a declaration that they understand this requirement and undertake to achieve it. On the other hand an experienced contractor may be unwilling to guarantee complete fulfilment of this obligation. This type of work is very difficult, and every effort should be made to anticipate potential problems before the contract is signed.

9.9.2 Repairs to honeycombed concrete

The location of the honeycombed concrete, i.e. whether it is on the inside face of the wall/floor/roof or whether it is deeper in the concrete member, will influence the method of repair. Generally, when the honeycombing is visible or very close to the inside surface, the repair can effectively be carried out as a surface repair. However, when the honeycombing is deeper in the concrete (and usually detected by an impulse radar survey or UPV), pressure grouting is usually better than hacking out and then repairing with concrete.

Both these repair methods have been described in Part 1 of this chapter, under section 9.4.5 to which the reader is referred.

9.9.3 Repairs to cracks and joints

Methods of repair to cracks and joints in concrete liquid retaining structures have been dealt with in Part 1 of this chapter, sections 9.4.2 and 9.4.3. The same principles apply to water-excluding structures, but renewal of sealants and repairs to concrete may have to be made against water pressure. This requires the inflow of water to be sealed off first by means of ultra-rapid setting proprietary compounds. When infiltration is considerable, a method sometimes adopted is to drill holes through the wall to provide additional outlet points and thus reduce the pressure of water at the areas to be repaired. These pressure relief holes have to be plugged in due course. As with all aspects of this type of work, considerable experience and practical knowhow is required.

9.10 CHLORIDE INDUCED CORROSION OF REBARS

9.10.1 General considerations

The investigation may have found that reinforcement corrosion has resulted from high concentrations of chlorides in the concrete. It has been emphasized previously in this book that with chloride corrosion of the rebars, normal repairs to the concrete including cleaning of rust from the rebars will only be effective for a relatively short time, probably 2–3 years. Therefore a more permanent repair method has to be used.

9.10.2 Cathodic protection

The long-term solution is cathodic protection as this system can deal with chlorides originally in the mix (and therefore evenly distributed throughout the concrete) and/or with chlorides penetrating the concrete from an outside source.

This system is now reasonably well established in the UK and has been briefly described in Chapter 6, section 6.6, to which the reader is referred.

9.11 OTHER REMEDIAL MEASURES: BASEMENTS

There are a number of items of work, in addition to those mentioned above, which can improve damp conditions in a basement and these are described below.

9.11.1 Control of ground water level

The ground water level can be lowered by provision of a perimeter drain discharging to a sump containing a pump and float control. The trench should be back-filled with hard core to facilitate sub-soil drainage. However, the possible effect of such a drainage system on the foundations of the building must be carefully considered, particularly in clay subsoils. All clays shrink when they dry out and some clays possess high shrinkage characteristics. The installation of sub-soil drainage around an existing building is likely to result in the reduction of the moisture content of the sub-soil below the foundations and this can cause foundation settlement in the long term.

9.11.2 Grouting the sub-soil for ground water control

In the early part of the century, grouts for injecting into the sub-soil were cement-based with the addition of selected clays. This was followed by chemical grouts which would gel, for example, sodium silicate with calcium chloride. The objective is to fill the voids in the sub-soil with the gel and thus effect a significant reduction in ground water flow through the grouted barrier. Grouting can be suitable for cohesionless soils, i.e. soils with a particle size in excess of about 0.002 mm (2 microns). Clay forms a major constituent of many chemical grouts. It is a complex material and its important characteristics depend on the clay minerals. For successful grouting, the calcium and sodium montmorillonites are particularly useful as they can produce gels which fill the subsoil voids and thus reduce the flow of ground water.

For successful ground water control by grouting, detailed information is required on the particle size and particle size distribution of the sub-soil as these control its porosity and permeability. Another important factor is the pH of the grout which should be alkaline. A type of clay known as Bentonite, which contains a high percentage of montmorillonite, has a pH of about 9.0 when dispersed in water.

The chemical grout curtain formed in the subsoil will greatly reduce the flow of ground water but is unlikely to form a complete cut-off. Repairs to the walls and floor of the basement or other water excluding structure will also be needed to ensure a complete repair. However, when the structure is a large one and serious infiltration is occurring, the use of sub-soil grouting may be essential for a successful job.

This work is highly specialized and should only be entrusted to contractors with the necessary experience.

9.11.3 Improvements to floor drainage

For basements of Grade 1 use, e.g. car parks, some form of floor drainage is always necessary. If the ingress of ground water is more than can be tolerated, it may be economic to improve the existing drainage system instead of carrying out large-scale repairs to seal off the leaks.

9.11.4 Control of vapour transmission

Basement usage of Grade 2 in Table 9.1 states that moisture vapour is tolerable. It is necessary to decide what level of relative humidity is tolerable for the basement use or proposed use; guidance on this can be obtained from BS 5250: Code of Practice for control of condensation in buildings.

The moisture level in the air is measured as relative humidity (RH). With the amount of water vapour constant, the RH will vary with the temperature, the lower the temperature the higher the RH until condensation occurs at the dew point. An engineer dealing with a damp basement should consult the appropriate authority, the Health and Safety Executive for factories and workshops, and the Environmental Health Officer of the local authority for shops and similar retail premises.

A generally accepted figure for RH for external air in the UK is 60–65%.

Basement usage Grades 3 and 4 can create considerable problems in achieving a satisfactory RH, particularly with a Grade 4 use. In addition to complete sealing of all leaks, the provision of a vapour-resistant membrane is required. With an existing basement this would have to be provided on the inside face of the walls and on the top surface of the floor slab. Such membranes generally consist of liquid polymer compounds applied in situ or polymeric sheet material.

For satisfactory performance, liquid membranes must bond well to the substrate and this requires proper preparation of the surface which must be free from dirt, grease, dust and other contamination. These materials are proprietary and the application of the suppliers should be followed.

Sheet material should be fully bonded to the substrate and then an inner skin of blocks (for the walls) and in situ concrete (for the floor) has to be provided.

For Grades 3 and 4 it is most likely that a properly designed system of ventilation and temperature and humidity control would be needed for commercial usage such as restaurants, shops etc. in addition to sealing off all leaks and the provision of a vapour-resistant membrane in the floor and walls.

9.12 PEDESTRIAN SUBWAYS

In addition to defects caused by honeycombed concrete, cracking and defective joints in the concrete retaining walls, I have found that trouble sometimes arises in this type of substructure, mainly in the walls, when they are finished in ceramic tiles or mosaic. The damage takes the form of the cracking and debonding of the finish. While this defect can occur at any location in the subway, it is more often found in the unroofed approaches (entrance and exit).

This damage is usually found to be due to three causes:

1. crack formation in the concrete retaining wall being reflected through the tile bed and tiles;
2. joints in the concrete wall not carried through the tile bed and tiles;
3. moisture and/or vapour pressure build-up behind the tiling.

Damage caused by (1) and (2) above can be readily detected by the removal of tiles and tile bed (and rendering if this has been applied to provide an even surface on which to bed the tiles).

Damage arising from (3) causing loss of bond and consequent displacement of tiles is not at all obvious and the cause can only be arrived at by careful investigation.

If investigation of the retaining wall design shows that a membrane was not provided on the earth side of the wall, and the tiles and tile bed have been properly laid and precautions taken to secure good physical bond between the various layers, it is reasonable to assume that vapour pressure has contributed to the loss of bond at the interface of the tiles and tile bed. The concrete wall, the rendering (if provided) and the tile bed are all to some extent porous, while the ceramic tiles themselves if they are frost resistant (vitrified) are almost impervious to the diffusion of moisture and water vapour.

With a tiled finish, the only exit for moisture and vapour is through the joints which are comparatively narrow (about 3–5 mm wide). As stated above, debonding appears to occur more frequently at the unroofed approaches to the subway, and this is not surprising as the walls are exposed to greater temperature changes and the sun shining on the walls will tend to draw out moisture and water vapour, causing a build-up behind the tiling.

For repair, a practical solution is to remove debonded and hollow-sounding tiles (and bed and rendering if this is also debonded) and refix using a styrene butadiene (SB) latex-based liquid membrane at the interface where loss of bond has occurred.

The SB latex based membrane should be applied to a clean slightly damp surface to give a minimum dry thickness of 0.75 mm, (approximately 1.4 mm wet thickness).

Repairs to cracks in the concrete retaining wall should be dealt with as previously described, with the addition of removal of all finishing layers for a distance of at least one tile width each side of the crack.

My experience is that cracks in reinforced concrete walls are more or less vertical and fairly straight. Therefore if the tiles are refixed with a vertical joint as close to the centre line of the crack as possible, the chances of new cracks forming in the tiling will be much reduced.

Damage caused by joints in the concrete not carried through the tiling indicates that the joints are 'live' and therefore to reduce the possibility of further trouble, joints must be formed in the tiling, tile bed and render, in line with the joints in the base concrete. This is unlikely to improve the appearance of the tiling, but would be better than cracked and debonded tiles.

PART 3 REPAIRS TO CONCRETE MARINE STRUCTURES

9.13 INTRODUCTION

Marine structures are located in a hostile environment, which is classified as 'most severe' in Table 5 of BS 5328: Concrete: Part 1. The description is:

> Concrete surfaces frequently exposed to sea water spray ... Concrete in sea water tidal zone down to 1.00 m below lowest low water.

It is therefore not surprising that experience shows that such structures are more liable to damage and deterioration than most land-based structures.

Technical publications on marine structures indicate the wide range of use to which such structures are put and this is illustrated in the following list:

- sea walls (quay walls, promenade and shore protection walls);
- jetties;
- dry docks;
- slipways;
- breakwaters;
- offshore structures for the petroleum industry.

BS 6349 Code of Practice for Maritime Structures is in seven Parts and Part 1 gives the following as reasonable estimates of design life:

Quay walls	60 years
Jetties	45 years
Shore protection works	60 years

9.14 CONSIDERATION OF THE PROBLEMS

Marine environment is a very wide term, as the location of the structures can vary from the tropics to the Arctic/Antarctic, and the detailed exposure conditions vary equally widely.

In the space of one chapter, I can only discuss the basic principles involved in the repair of structures in a marine environment.

9.14.1 The salts (sulphates and chlorides) in sea water

The one common factor in the wide range of environmental conditions is that the structures are in constant or intermittent contact or immersion in a relatively high concentration of dissolved salts, mainly chlorides and sulphates. The figures in Table 9.2 are given by Lea in *The Chemistry of Cement and Concrete*, 3rd edition, p. 65.

Table 9.2

	Atlantic	*Mediterranean*
Chlorides	17.83	21.38
Sulphate (as SO_4)	2.54	3.06

The above figures are in grammes per litre; converted to ppm or milligrammes per litre, they approximate to the figures given in Table 9.3.

Table 9.3

	Atlantic	*Mediterranean*
Chlorides	18000	21000
Sulphate as SO_4	2500	3000

It should be noted that the figure for sulphates given above is expressed as SO_4 whereas in Standards and Codes sulphate is usually expressed as SO_3. To convert SO_4 to SO_3:

$$SO_3 = SO_4 \times 80/96 = 0.83 \times SO_4$$

The calculation is based on atomic weights: sulphur = 32 and oxygen = 16.

$$SO_3 = 32 + 3 \times 16 = 80; SO_4 = 32 + 4 \times 16 = 96.$$

From the above it can be seen that sulphate as SO_3 in Atlantic water is approximately 2100 ppm and in Mediterranean water is approximately 2500 ppm. Both these sea waters would fall within Class 3 ground water for the purpose of assessing possible sulphate attack on the concrete (see BS 5328: Part 1, Tables 7a and 7b. Therefore according to the Standard, ordinary Portland cement on its own would not comply with the Standard.

I have not seen any authoritative figures for the sea water in the Arabian Gulf, although a great deal of construction work has been carried out there during the past 40 years. At the time I worked in Kuwait, the following figures given in Table 9.4 were generally accepted.

Table 9.4

	Arabian Gulf
Chlorides	25000 ppm
Sulphate (as SO_3)	3000 ppm

It is known that a very large number of marine structures have been constructed in Atlantic water with OPC without suffering sulphate attack. The quality of the concrete used for marine structures is very high, and for reinforced concrete a maximum water/cement ratio of 0.45, a minimum cement content of 400 kg/m^3 and a characteristic compressive strength of 50 N/mm^2 is normally specified. Concrete of this quality ensures its resistance to sulphate attack when made with OPC, unless sulphate concentration is significantly higher than that found in sea water.

9.14.2 Chlorides from sea water in concrete

A major factor is the presence of chlorides in concentration of 18 000 ppm, 21 000 ppm and 25 000 ppm. The presence of chloride ions in the concrete can stimulate corrosion of the steel reinforcement even when the alkalinity (pH) is high. The effect of the chloride ions is complex as they combine with the tricalcium aluminate (C3A) in the cement to form compounds which effectively prevent the chloride ions from attacking the steel. The higher the percentage of C3A the more chloride ions will be immobilized.

Sulphate resisting Portland cement has a low C3A content; BS 4027 limits the C3A content to 3%, while OPC is likely to contain 6% to 12% C3A.

I believe that it is better to use a cement with a high C3A content than a sulphate resisting cement with a low C3A content, unless there are special reasons not to do so.

It has been suggested that concrete containing either ground granulated blast-furnace slag (GGBS) – BS 6699, or pulverized fuel ash (pfa) – BS 3892) will provide better protection to the rebars against chloride attack than OPC or SRPC.

As far as I am aware, this is still open to argument by some experienced engineers.

9.15 CAUSES OF DETERIORATION

Damage and deterioration can occur to a marine structure in many ways which can be considered under three main headings:

1. physical damage;
2. chemical attack;
3. a combination of (1) and (2).

Each of the above main categories can arise from a number of causes and when considering repairs the real cause(s) of the problem must be determined before a satisfactory method of repair can be devised.

9.15.1 Physical damage

The type and general cause of physical damage will depend mainly on the type of structure, its use and its location, and this is summarized below.

1. In the Arctic regions damage by ice is quite common and this has to be taken into account.
2. Exposure to freeze–thaw conditions, in and above the splash zone in certain climatic conditions.
3. Exposure to wave action.
4. Jetties and quay walls are likely to suffer damage from vessels berthing alongside.
5. Jetties and similar which form part of an oil terminal may be seriously damaged by a hydrocarbon fire. Figure 9.8 shows one small section of what was overall very extensive damage; the damage to the concrete extended for more than 300 mm behind the 32 mm dia. reinforcing bars.
6. Promenade walls and walls constructed to prevent sea encroachment are likely to suffer damage and abrasion from sand and

Figure 9.8 Serious damage to concrete jetty by hydro-carbon fire. Courtesy, Burks Green & Partners Consulting Engineers.

shingle being thrown against the concrete in times of storm. Figure 9.9 shows severe abrasion of the base of a concrete pier caused by sand and shingle.

7. Another form of physical damage occurs to concrete in the splash zone (from lowest low water level to a variable height above highest tide level). This is caused by the crystallization of salts in solution which have been absorbed into the surface layers of the exposed concrete.

The formation and growth of crystals in the surface layers of the concrete results in shallow spalling and pitting, which weakens the surface and allows further ingress to take place. Concrete, unlike metals and plastics, is a porous material because it possesses a pore structure. The pores are very small and, according to F.M. Lea, the distinction between gel pores and capillary pores in the hydrating cement paste is rather artificial and the real distinction is between small and large pores which overlap in their range.

9.15.2 Corrosion of steel reinforcement

An important type of physical damage to reinforced concrete in a marine environment is cracking and spalling caused by the corrosion/rusting of the rebars. When steel rusts, the corrosion product

Figure 9.9 View of abrasion of concrete pier by shingle and sand.

occupies a volume about 3–5 times the volume of original steel and this expansion disrupts the concrete. The principle cause of the corrosion of the rebars is the ingress of sea water containing chloride ions, provided oxygen and moisture are present. The availability of oxygen decreases with increasing depth below low water level.

In the 1970s a major research programme was sponsored by the Department of Energy and named 'Concrete in the Oceans'. The aim was to provide knowledge on the long-term performance of reinforced concrete oil production platforms, mainly in the North Sea, anchored into the sea bed at depths down to 250 m.

I have found that three of these Reports contain information of particular relevance to problems of deterioration of normal type marine structures. The Reports in question are:

Report No. 2/11: 1976	*Cracking and Corrosion*, by A.W. Beeby
Report No. 5: 1980	*Marine Durability Survey of the Tongues Sands Tower*, Taylor Woodrow Research Laboratories
Report No. 6: 1980	*Fundamental Mechanisms of Corrosion of Steel Reinforcement in Concrete immersed in Sea Water*, by N.J.M. Wilkins and P.F. Lawrence

The Reports confirmed that the most severe conditions from the point of view of deterioration occur in the splash zone. Also, that rebars embedded in concrete which is permanently submerged at a reasonable depth. This is due to the substantial reduction in available oxygen.

The passivity provided by the cement paste in the concrete surrounding the rebars is only reduced slowly, and thereafter active corrosion proceeds very slowly compared with concrete containing similar concentration of chlorides exposed to the air (the splash zone and to some extent above this zone).

Where corrosion is caused by chlorides in the concrete which is in contact with the rebars, localized (pitting) corrosion is likely to occur and this can cause more serious damage to the rebars than general corrosion. The pitting may penetrate the steel to more than 50% of the bar diameter.

9.15.3 Chemical attack

I have not seen any reports on the deterioration of marine structures located in what may be termed the open sea caused by chemical attack on the concrete. However, I have dealt with a case of chemical attack on a structure located in a tidal estuary.

Chemical analysis of the damaged concrete showed that sulphate attack had not taken place. Examination of the concrete clearly indicated attack, probably by industrial effluent. It was concluded that the attack had taken place intermittently over many years. At the time of the investigation analysis of the estuary did not reveal aggressive chemicals in sufficiently high concentration to cause the damage found. This damage occurred below low water level and within the tidal zone.

9.16 INVESTIGATIONS

While it is probably true to say that 90% of the damage to reinforced concrete marine structures arises directly or indirectly from the corrosion of the steel rebars, a careful investigation is always desirable.

For example, a marine structure in the Mediterranean showed signs of serious deterioration in less than ten years after completion. An investigation showed that the main cause was alkali–aggregate reaction. This was entirely unexpected as there was no record of previous cases although considerable use was made of concrete for land-based buildings.

In the Arabian Gulf most concreting aggregates were found to be contaminated with chlorides and sulphates and some of the mixing water was to some extent saline. A changeover to the use of sulphate resisting Portland cement did not result in any improvement – rather the reverse; the reasons for this are given in section 9.14.2.

The basic principles described in Chapter 4 should in general be followed. However, carbonation of the concrete is unlikely to play a part in the corrosion of the rebars except in a small number of members well outside the spray zone. The following factors should be investigated:

1. the original design of the structure and any major alterations since;
2. climatic variations;
3. exposure conditions relevant to direction of storm tracks;
4. wave characteristics;
5. tide range;
6. any adverse effect on the concrete by marine life;
7. shipping data and berthing procedure (if applicable);
8. if water is discharged from the structure, such as cooling water from a power station, its temperature and chemical characteristics;
9. chemical characteristics of the sea water in which the structure is built including any trade effluents.

For the investigation of damage or possible damage below low water level, closed-circuit television can be very useful for providing information for an initial appraisal of the situation. For a more detailed investigation an examination by divers is usually necessary.

Existing structures are usually covered with marine growth, such as seaweed, barnacles etc. and these will have to be removed in order to ascertain the extent of the damage to the concrete. Marine growths of all types can be readily removed by high-velocity water jets, operating with nozzle pressures of about 400 atm.

9.17 METHODS OF REPAIR

9.17.1 General

The **principles** of repair described in Chapters 5 and 6 for land-based structures apply to marine structures, but **methods** of repair are likely to be different and be greatly influenced by the information obtained from the investigations outlined in the previous section.

The extent of the deterioration is usually much more serious and the repaired areas have to stand up to a very hostile environment. The conditions under which the repair work has to be carried out are usually very difficult. Repairs within the tide range have to be carried out to a limited time schedule and work below low-water level presents many problems.

The majority of repairs to the structure above high tide level are likely to come under the general heading of 'patch' repairs.

The use of a high-quality coating has been recommended in Chapters 5 and 6, and some detailed information on suitable coatings is covered in Chapter 7. Even the best of coatings may not provide good long-term performance in a marine environment. Whenever possible, I recommend the application of gunite over the whole surface of all repaired members, to a minimum thickness of 50 mm, reinforced with a galvanized steel mesh securely fixed to the base concrete. 'Gunite' is also known as pneumatically applied mortar, sometimes referred to as shotcrete, and if coarse aggregate (10 mm and above) is used it is known as sprayed concrete. The effect of the increase in dead load should be checked.

The above brief outline of 'patch' repairs does not take into account the almost certain presence of concentrations of chloride ions in contact with the rebars. The only practical and effective way of dealing with this situation is to introduce a system of cathodic protection, which is discussed in the next section.

Generally, the most difficult type of repair to a marine structure is when this lies within the tide zone or below low water level.

The actual method used for the repair will depend on the circumstances of each case, and may call for the use of concrete placed by a tremie pipe or skip. Reference can be made to the Concrete Society Technical Report TR 035 – Underwater Concreting.

Special mix design is called for including precautions to prevent segregation. With underwater concreting it is particularly difficult to ensure good bond between the existing concrete and newly placed concrete.

Good results have been reported on the use a technique which involves filling the formwork with graded aggregate and then injecting a specially formulated cement-based grout.

Underwater repairs is a job for the specialist and should only be entrusted to firms which can demonstrate previous successful performance. The checking of such claims can itself be difficult as the actual repairs are under water and cannot be readily inspected.

It is reasonable to assume that responsible firms offering special repair systems will have carried out tests before the technique was originally used. Details of such tests should be available on request. This is normal practice for land-based structures and often includes the grant of a Certificate by the British Board of Agrement.

9.17.2 Cathodic protection

General information on cathodic protection has been given in Chapter 6, section 6.6, but some additional comments are considered desirable in the case of marine structures.

Due to the inevitable diffusion of chloride into the concrete in marine structures, the corrosion of carbon steel rebars is bound to occur. Whether this corrosion starts soon after construction or some years later depends on a number of factors. The principal ones are:

1. permeability of the concrete, which largely determines the rate at which chloride ions diffuse into the concrete and reach the rebars and destroy the passivation provided by the cement paste;
2. the depth below low water level at which the repair has to be carried out; the greater the depth the slower will be the rate of corrosion due to reduction in available oxygen.

As far as I am aware, the only practical way to stop this corrosion is by the installation of a properly designed system of cathodic protection.

The cost of the installation of a comprehensive system of cathodic protection is high, probably four times that of 'patch' repairs, which in a marine structure are likely to become a continuous operation as more and more steel becomes corroded and cracks and spalls the concrete.

The two basic systems of cathodic protection are sacrificial anodes and impressed dc current, the latter being the system usually adopted.

Further information on the use of cathodic protection is given in the papers listed on in the 'Further reading' section that follows.

9.18 FURTHER READING

Repairs to concrete liquid-retaining structures

British Standards Institution (1987) *Code of Practice for Design of Concrete Structures for Retaining Aqueous Liquids*, BS 8007.
British Standards Institution (1991) Cathodic Protection, BS 7361: Part 1: 1991, *Code of Practice for Land and Marine Applications*.
Bamforth, P. (1992) The development of cathodic protection systems, *Concrete*, Mar./April, 48–52.
Construction Industry Research and Information Assoc. (1987) *Civil Engineering Sealants in Wet Conditions*, Tech. Note 128, 76.
Construction Industry Research and Information Assoc. (1991) *Water-proofing Reservoir Roofs*, Tech. Note 145, 103.
Green, J.K. and Perkins, P.H. (1980) *Concrete Liquid Retaining Structures*, Applied Science Publishers, 355.
Hammond, A.D. (1993) Electrochemical techniques for repair of concrete, *Construction and Repair*, July/Aug., 8–12.
Perkins, P.H. (1981) Corrosion problems in sewerage structures, Paper No. 12, *Proc. Inter. Conf. Inst. Civ. Engr.*, London, June, 133–9.
Thistlethwaite, K.D.B. (1972) *Control of Sulphides in Sewerage Systems*, Butterworth, Sydney, Australia, 273.

Repairs to concrete water-excluding structures

Bowen, R. (1975) *Grouting in Engineering Practice*, Applied Science Pub. Ltd, London, 187.
British Standards Institution (1990) BS 8102: 1990: *Code of Practice for Protection of Structures against Water from the Ground*.
British Standards Institution (1989) BS 5250: 1989: *Code of Practice for Control of Condensation in Buildings*.
Building Research Establishment (1972) Condensation, *Digest*, **110**, 8.
Building Research Establishment (1985) *Mould and its Control*, IP, 11/85 June, 4.
British Cement Assoc. (1994) *Basement waterproofing – design guide*, 48.058, 19.
British Cement Assoc. (1994) Basement water proofing – site guide, 48.059. 21.
Colvil, C.S. and Skinner, A.E. (1992) Jet grouting – a review, *ICE. Proc. of Conf.* London, Nov.
Concrete (1990) Review of waterproofing systems, June, 33–7.
Construction Industry Research and Information Assoc. (1995) *Water Resisting Basement Construction – A Guide*, Report, 139, 187.
Dibb-Fuller, E. (1990) 60 Victoria Embankment: profile of a deep basement construction, *Concrete*, **24**(6), June 27–32.
Evans, P.B. (1995) Damp? Not in my book, *Concrete*, Jan/Feb. 47–8.
Leek, D., Johnson, A. and Cope, M., (1995) How waterproof is your basement?, *Concrete*, Jan./Feb., 20–3.
Morgan, J.P. (1990) 60 Victoria Embankment: profile of a deep basement construction, *Concrete*, June, 27–32.

Repairs to concrete marine structures

ASTM (1984) *Standard Specification for Epoxy-coated Reinforcement Steel Bars*, ASTM/A775M-84.

BSI (1990) *Fusion Bonded Epoxy-coated Carbon Steel Bars for the Reinforcement of Concrete,* Parts 1 and 2.
BSI (1984–1991) *Maritime Structures,* BS 6349, Parts 1–7: 1984–1991.
BSI (1991) *Cathodic Protection:* Part 1: 1991: *Code of Practice for Land and Marine Applications.*
Bamforth, P. (1994) Reinforcement in marine structures, *Concrete,* Jan./Feb., 33–6.
Billington, S.J. (1979) The underwater repair of concrete off-shore structures, *Off-Shore Technical Conference,* Houston, Texas, April/May, Ref. OTC 3464, 927–37.
Concrete Society (1986) *Concrete in the Marine Environment,* 30 papers presented at the Marine Concrete International Conf., London, Sept.
Concrete Society (1990) *Underwater Concreting,* Ref. TR 035, 48.
Concrete Society (1989) *Cathodic Protection of Reinforced Concrete,* Ref.TR 036, 60.
Foukes, P.G. (1993) Concrete in the Middle East, past, present and future: a brief review, *Concrete,* July/Aug., 14–20.
Gjorv, O.E. (1971) Long-time durability of concrete in sea water, *ACI Journal,* Jan. Title 68–10, 60–7.
McAnoy, R.P.L., Palmer, R.A. and Longford, P.E. (1987) Cathodic protection of reinforced concrete structures – experience in UK, Hong Kong and Australia, *2nd International Conference, Deterioration & repair of reinforced concrete in the Arabian Gulf,* Oct. Bahrain, 121–38.
Treadaway, K.W.J.A. (1988) Method of Evaluation of Repairs to Reinforced Concrete in Marine Conditions, BRE Information Paper, IP 11/88, Oct., 3.
Young, C. (1994) Repair of marine structures in the Gulf, *Construction Repair,* July/Aug., 35–9.

Index

Page numbers in **bold** refer to figures; those in *italic* refer to tables.

For Product Safety Concerns and Information please contact our EU
representative GPSR@taylorandfrancis.com
Taylor & Francis Verlag GmbH, Kaufingerstraße 24, 80331 München, Germany

www.ingramcontent.com/pod-product-compliance
Ingram Content Group UK Ltd.
Pitfield, Milton Keynes, MK11 3LW, UK
UKHW022001180926
475701UK00003B/130
* 9 7 8 0 3 6 7 8 6 6 3 9 6 *